THE CAREY ACT AND CONSERVATION IN COLORADO

THE CAREY ACT AND CONSERVATION IN COLORADO

Gerald C. Morton

UNIVERSITY PRESS OF COLORADO
Denver

Published by University Press of Colorado
1580 North Logan Street, Suite 660
PMB 39883
Denver, Colorado 80203-1942

Printed in the United States of America

The University Press of Colorado is a proud member of Association of University Presses.

The University Press of Colorado is a cooperative publishing enterprise supported, in part, by Adams State University, Colorado State University, Fort Lewis College, Metropolitan State University of Denver, University of Alaska Fairbanks, University of Colorado, University of Denver, University of Northern Colorado, University of Wyoming, Utah State University, and Western Colorado University.

∞ This paper meets the requirements of the ANSI/NISO Z39.48-1992 (Permanence of Paper).

ISBN: 978-1-64642-648-5 (hardcover)
ISBN: 978-1-64642-649-2 (ebook)
https://doi.org/10.5876/9781646426492

Library of Congress Cataloging-in-Publication Data

Names: Morton, Gerald C. (Gerald Charles), author.
Title: The Carey Act and conservation in Colorado / Gerald C. Morton.
Description: Denver : University Press of Colorado, [2024] | Includes bibliographical references and index.
Identifiers: LCCN 2024008374 (print) | LCCN 2024008375 (ebook) | ISBN 9781646426485 (hardcover) | ISBN 9781646426492 (ebook)
Subjects: LCSH: Carey, Joseph M. (Joseph Maull), 1845–1924—Political and social views. | Irrigation projects—Colorado—History. | Reclamation of land—Colorado—History. | Agricultural development projects—Colorado—History. | Environmental protection—Colorado—History. | Wildlife conservation—Colorado—History. | Irrigation laws—Colorado—History. | Arid regions—Colorado—History. | Public lands—Colorado.
Classification: LCC HD1739.C6 M67 2024 (print) | LCC HD1739.C6 (ebook) | DDC 338.109788/99—dc23/eng/20240612
LC record available at https://lccn.loc.gov/2024008374
LC ebook record available at https://lccn.loc.gov/2024008375

Cover photograph of Two Buttes Reservoir by Diane Roberts

For Diana

CONTENTS

ILLUSTRATIONS

FIGURES

MAPS

ACKNOWLEDGMENTS

The Carey Act has long been an obscure chapter in Colorado history. It was a reclamation policy failure, a significant one. Perhaps the reason for that obscurity resides in our tendency to forget disappointment. Yet remarkably, from its failure came successes—wildlife areas that today define an arid landscape where irrigation development could not.

My interest in the Carey Act comes from a lingering curiosity I have had about my family's connection with the law's failure in Colorado. But it also comes from years of appreciation I have had for the conservation of wildlife areas, especially those that were once Carey Act projects in southeastern Colorado. My maternal great-grandfather, George E. O'Brien, headed the Las Animas–Bent Carey Act Project twenty miles south of Las Animas, Colorado, where he and other family members settled after World War I. The project's abysmal failure drove the pioneering family off the land, and the episode has long been something of a family embarrassment. Regrettably, his records about the development were destroyed in a farmhouse fire in 1931. Nevertheless, a wealth of historical records exists about the Carey Act, especially its application in Colorado. As I uncovered my family's role in the unfortunate and tangled affair, I began to see a larger story of the Carey Act's consequence in the history of Colorado and the American West. Here was a story worthy of a book about why these western developers went to such inexorable lengths in trying to build unviable project after unviable project. Here too was a deep layer of tension between developers and settlers that told a gripping story. Moreover, the Carey Act's conservation aftermath gives this forgotten story a modern-day relevance that links the past to the present.

As an avid hiker, I made it a point to explore each of the benchland locations where Carey Act developers desperately believed it was possible to transform arid stretches into irrigated farmsteads. A total of thirty-four Carey Act projects were scattered across the state. As I walked the land, I was struck time and again by the physiographic challenges the benchlands presented to developers and settlers. Extremely arid conditions define each location. Obviously apparent was how limited the water supply was. The intermittent nature of supplying streams and a dependence on floodwaters were difficulties that doomed the Carey Act in Colorado. But those environmental conditions proved ideal habitat for the restoration of native wildlife and the introduction of some non-native species. I thank all the landowners and public land agencies for granting me permission to examine their properties so closely and to witness the abundance of wildlife I encountered along the way.

My dive into the historical record of the Carey Act in Colorado has taken me to the far reaches of the state as well. Archivists, librarians, and curators—Clio's stewards—have been collecting and caring for documents about the Carey Act for more than a century, all the while preserving documents about the development of wildlife conservation. Today's record keepers have a remarkable command of their collections. I owe each of the following individuals a sincere thank-you for locating all the ledgers, files, and photographs that comprise the foundation of this book: Erin McDanal, Lance Christensen, Elena Cline, and Annie Epperson, Colorado State Archives; Cindy Brown, Wyoming State Archives; Mark Anderson, University of Northern Colorado; Andy Senti, Bureau of Land Management, Denver; Caroline Blackburn and Melissa Gurney, Miranda Todd, and Katie Ross, City of the Greeley Museum; Patti Rettig, Archives and Special Collections, Colorado State University; Bill McCormick and Mark Perry, State Engineer's Office, Pueblo; John Clark and Travis Black, Colorado Parks and Wildlife; Dan Davidson and Janet Gerber, Museum of Northwest Colorado; Donna Dodson and Kathleen Tomlin, Bent County Historical Society; and Theresa Hendricks, former curator of the Doc Verity Museum, Two Buttes, Colorado.

Research and writing can be a lonely business at times. I am especially thankful for the counsel I received from the following people who went above and beyond the call of duty in helping me with this book. Professor Donald J. Pisani encouraged me to pursue a study of the Carey Act in Colorado. Two anonymous reviewers for the University Press of

Colorado deserve enormous thanks for providing invaluable assistance that included ways I might rethink my argument and refine my manuscript. To the editors at UPC over the years—Jessica d'Arbonne, who saw a diamond in the rough, as well as Allegra Martschenko, Robert Ramaswamy, and Laura Furney for welcoming an independent historian; to Tina Kachele and Daniel Pratt for their attention to the page layout and cover; and to my copyeditor, Cheryl Carnahan, I thank you so very much. Finally, I would be remiss if I did not acknowledge the many environmental historians whose writing influenced my thinking about human interaction with nature. Their scholarly contributions, mentioned throughout this text and in the bibliography, contributed immensely to this book.

THE CAREY ACT AND CONSERVATION IN COLORADO

INTRODUCTION

The Purgatoire River, southeastern Colorado's longest and largest watercourse south of the Arkansas River, originates in the Sangre de Cristo Mountains and flows east to the plains. There, for seventy-five miles it cuts through a slightly uplifted land of folds and domes before emptying into the Arkansas. The region is an ecological wonder of highlands, canyons, juniper-covered hills, and grasslands that form a dendritic network of ephemeral tributaries that rush occasional floodwaters into the Purgatoire. Some of those dry or nearly dry streams extend as far east as the low shelf lands of the Hugoton Embayment near the Kansas border before emptying into the Arkansas.

Remarkably, for at least 100 centuries, Native Americans occupied this land, living with nature's limits and excesses before the United States dispossessed them of title to it by the 1860s. Hispanic and white settlers began filing for title here in the mid-nineteenth century, the former fashioning livelihoods by simple irrigation subsistence farming and the latter by grazing livestock and dry farming. Into this setting, along two of the

https://doi.org/10.5876/9781646426492.c000

region's nearly dry streams, determined irrigation developers intervened with nature and brought to fruition Colorado's only reclamation projects constructed under the United States Federal Desert Land Act of 1894, more commonly called the Carey Act after its sponsor, US senator Joseph M. Carey of Wyoming. The law offered each western state 1 million acres of desert-classified federal land for reclamation by development companies to spur private irrigation and the settlement of actual settlers. The Two Buttes Project in Baca County drew a colony of 200 settlers in 1911, and the Muddy Creek Project in Bent County lured nearly as many settlers in 1919. Here, in a region that receives less than fifteen inches of annual precipitation, the developers time and again went to extraordinary means in an effort to create irrigated paradises that were akin to nearby irrigation development in the well-watered valleys of the upper Purgatoire and the lower Arkansas. In their zeal, the developers built extensive waterworks to capture floodwaters, only to experience financial failure, leaving settlers to fend for themselves. By 1970 the descendants of these Carey Act settlers sold the last of the private waterworks to the State of Colorado, which for years had managed parts of Two Buttes and Muddy Creek Reservoirs as wildlife conservation areas. The state continues to administer both public wildlife areas in perpetuity through its Parks and Wildlife Department, and nature's limits and excesses likewise continue to dictate the agency's charge of wildlife conservation.

This book explains the nature of the Carey Act and its application in Colorado; it concludes with why Two Buttes, Muddy Creek, and several other scattered remnants of the 1894 US law's failure have become important wildlife conservation areas. It argues that Carey Act developers, few of whom had experience in agriculture, refused to accept the limitations of their schemes but persisted nonetheless, often until well after financial ruin eventually crushed their hopes of building vibrant and enduring communities. The values of the developers seeking windfall profits were at constant odds with the values of settlers seeking viable livelihoods. That conflict explains much about the Carey Act's failure in the Centennial State. In turn, from that spectacular failure, various conservation-minded individuals and organizations advocating the sweeping ethic of environmentalism seized the opportunity to transform the most developed of the schemes into publicly owned lands and waters. These places, although quite small—Two Buttes at 8,533 acres and Muddy Creek at just 2,438 acres—are unique habitats, set-asides for the exclusive benefit of wildlife. They are the earliest examples of wildlife conservation efforts on Colorado's southeastern plains. Moreover, they are the example from which more than 100,000 acres of subsequently established state wildlife areas across that horizontal landscape have blended into the farming- and grazing-based economy of the early twenty-first century.

The Two Buttes and Muddy Creek developments were isolated. Their contrast could not have been more dramatic from the developments of the upper Purgatoire Valley and especially along the lower Arkansas Valley, where irrigated bottomlands and easily irrigated benchlands numbered tens of thousands of acres. These farmers might put as much as 5 acre-feet of water on various crops per year, as opposed to 1.5 acre-feet at Two Buttes and Muddy Creek. From Pueblo to the Kansas state line, a string of railroad towns grew, and the sugar beet industry transformed the business of farming in 1900. The earliest significant irrigation development in Colorado had occurred along several of the state's rivers, where settlers had built self-funded irrigation ditches during the 1860s and 1870s. Sufficiently capitalized corporations later constructed larger irrigation systems and leased or sold land and water to farmers on adjacent benchlands. By 1900 many of these enterprises, lacking adequate water supply and a legal right to water during droughts, proved unprofitable; farmers had come to possess pieces of the systems reorganized as mutually owned irrigation companies. Notwithstanding shifts in the ownership of irrigation enterprises, Colorado, since statehood in 1876, bound the use of water to its water appropriation law that granted users the right to divert water in order of priority based on the earliest-in-time users.[1]

Carey Act developers nearly always looked to the very remote benchlands, and their water rights, which they sold appurtenant (attached) to the land, were always junior in priority. As the reader will see in the following chapters, each of their schemes to capture floodwaters proposed using just 1.5 acre-feet of water to irrigate high-profit crops such as alfalfa, corn, and sugar beets. Across Colorado they fantasized about building thirty-four new irrigation projects between 1902 and 1921 and proposed spending $30 million to make desert lands bloom. By 1925 every Carey Act project in Colorado was an abject failure, and the state effectively ceased

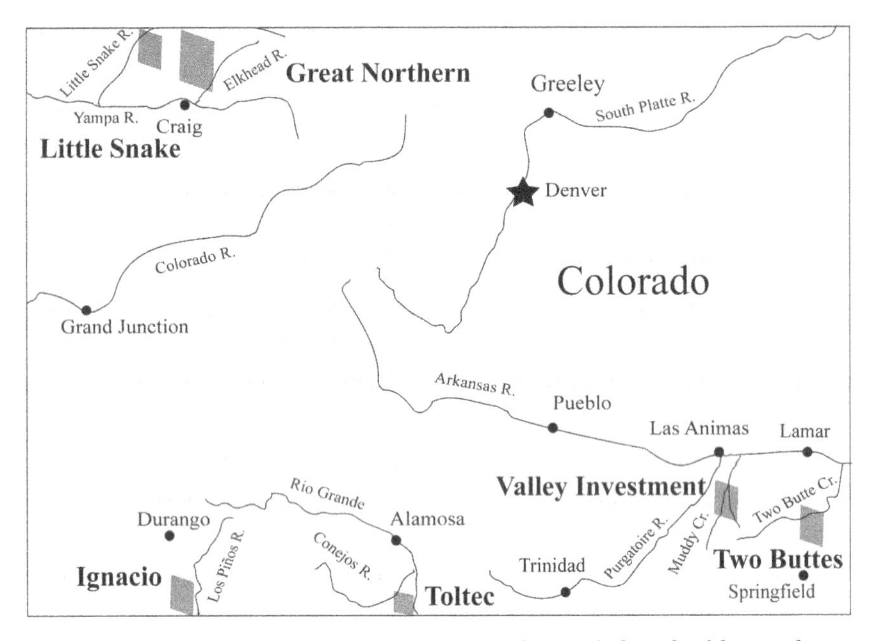

MAP 1. Carey Act projects in Colorado segregated, parceled, and sold to settlers

participating in the law's offering. In addition to the Two Buttes and the Muddy Creek Projects, only four other Carey Act developments advanced beyond the promotion stage to construct canals and begin locating settlers. In northwestern Colorado's Moffat and Routt Counties, developers undertook building the Little Snake River Project (38,000 acres) and the Great Northern Project (142,732 acres); in southwestern Colorado's La Plata County, the Ignacio Project (16,000 acres); and in the San Luis Valley in Conejos County, the Toltec Project (14,852 acres). These four projects, however, never constructed significant waterworks such as dams and advanced canal systems and thus only feature in this story of the Carey Act as they relate to the law in a general sense. A list of the Carey Act projects across Colorado, with details of each and a map of their location, as well as the full text of the Carey Act appear in the appendixes.

The Carey Act was an early public policy example of cooperative federalism. It offered a package deal for the land: free segregations of federal land to the states to attract heavy capital investment to construct substantial irrigation systems and charge settlers under state regulation from $35 to

$60 per acre—enough to cover costs and turn a substantial profit. Legally binding contracts bound each party to the deal. Once settlers had paid off the financing of their systems, they were to assume mutual ownership of the waterworks and title to their farms for which they had pledged, not mortgaged, to lenders. To pay for a state's expense in administering the law, including eventual certification of a farmer's land patent, the state charged the settler a nominal sum of fifty cents per acre. This was the law's rationale: that people might afford to settle in many arid regions, and the federal government would be free of having to pay the extraordinary cost of developing irrigation. To a settler, the law offered the acquisition of land with irrigation at a cost as little as one-fifth that of land along established canals.

Moreover, the developers' business was transactional, centered more on the exchange of money than on a lasting concern for the success of farmers. But as historian Robert Pisani has written, developers also emphasized, as did other irrigation advocates, the moral good in the economic dimension of their efforts and also the public good in the settlement of arid lands. They sold settlers the garden myth that American virtues derived from an agrarian way of life and that irrigation specifically offered farmers a way to avoid the ever-present likelihood of bankruptcy that dryland farming often brought. Irrigation, they said, also paid better because it increased yields, allowed for crop diversity, and might assure a high-value crop such as sugar beets or alfalfa even in dry years. In addition, clusters of small family farms under irrigation systems could create viable communities across the vast stretches of empty lands. There, American capitalism could flourish and strengthen the American family and the middle class.[2]

Colorado formally sought to claim less than 300,000 acres under the Carey Act. Patented (titled) land from the federal government to the State of Colorado amounted to a mere 37,302 acres: 13,302 acres at Two Buttes and 24,000 acres at Muddy Creek. Although developers took the four other projects to the point of settling farmers, only the two southeastern Colorado developments furnished water to the land. Nevertheless, the Two Buttes Project went bankrupt in 1927, though some of its colonists persisted through the Dust Bowl era and later periods of drought and grasshoppers, and irrigated from their reservoir until they sold it to the State of Colorado in 1970. The Muddy Creek Project collapsed entirely in 1945 after decades of failure. Today, Two Buttes Reservoir is the central feature of the Two Buttes State Wildlife Area. The Muddy Creek Reservoir site comprises the Setchfield State Wildlife Area, which Colorado purchased in 1956. A raging torrent in 1965 breached this dam, and wildlife officials manage it as a dry lake ecosystem.

Consequently, this book aims to serve two audiences. For the general reader with an interest in the history of the American West, it offers an inclusive narrative of the little-known Carey Act in Colorado. It explains the law's origins and its saga of repeated failures, the untold stories of developers and a few of the settlers who chanced irrigating from intermittent streams that flow less than five months annually or ephemeral streams that flow only during flash floods. And it explains the law's aftermath of conservation successes and challenges. For the environmental historian, it offers the example of how nature pushed back against capitalism and how nature dictates its own conservation design despite human efforts to conserve it. This book makes an important contribution to the study of agricultural development in the West by offering a counterpoint to prevailing studies of the Carey Act's relative success in Idaho, some projects in Wyoming, and one development in Utah. Those well-examined Carey Act projects are not analogous to the 1894 law's unfortunate saga in Colorado. Moreover, a key intervention this book makes is to place the story of the Carey Act's failure in Colorado at the center of wildlife habitat conservation efforts that led people to rethink and reconfigure their relationship with the arid land and capitalism's exploitation of it on the state's southeastern plains. Those restoration and preservation efforts, explained here for the first time, elevate wildlife conservation's importance into the greater discussion of soil conservation and water conservancy policies of federal, state, and local authorities during the post–Dust Bowl years. Hence, this book correlates patterns of reclamation failures decades in the making with conservation outcomes.[3]

Some readers might see in this agriculture-to-conservation conversion an inevitable progression, a teleological outcome that forces of nature and circumstance have guided—swallows nesting under a bridge, if you will. Indeed, tens of thousands of geese came to winter annually at Two Buttes Reservoir. And thousands still do. The endangered humpback chub today swims below a reservoir contemplated by the irrigation developers of the Great Northern Carey Act Project. Yet only because of the establishment of comprehensive wildlife conservation policy initiatives dating to the 1930s and vastly expanded across the state since then do the geese continue to light and does the once nearly extinct fish survive. Here is a story not of environmental and moral declination but of correcting errant steps.[4]

Those missteps for Colorado began with the creation of the Carey Act. As detailed in chapter 1, Senator Joseph M. Carey, policymakers who supported his 1894 law, and developers who chased its offering of windfall profits embraced the deeply held rationalizations of America's westward expansion. Economic and social justifications informed these men's thinking. As historians William Cronon and Richard White each emphasize, the capitalist market system, which commodifies nature, was an essential dynamic for development in the American West. Indeed, the Carey Act's design, similar to its unsuccessful predecessor policy proposal of arid land cessions to states, spurred the creation of wealth by making public domain private property, thus advancing American values. Moreover, the Carey Act was among the earliest of many public US policy proposals that sought to bend nature's design for material benefit across the arid regions of the West. The subsequent Reclamation Act (1902) set into motion the most consequential of all reclamation policies, a course that historian Donald Worster so aptly coined the Hydraulic Society.[5]

Chapter 2 traces the unsuccessful efforts of Greeley-area developers Daniel A. Camfield and George H. West, beginning in 1895, to utilize the Carey Act in Colorado along undeveloped benchlands adjacent to the South Platte River. It shows how the Carey Act's early bureaucratic inadequacies as well as the ongoing economic depression of 1893—the worst to that date in the nation's history—made marketplace financing impossible. That failure proved fortuitous because it allowed the ever-resourceful developers to utilize Colorado's 1905 Irrigation District Act and the convention of mutual irrigation companies to finally undertake their developments. As well examined by two separate historians, Daniel Tyler and Michael Weeks, those irrigation institutions paved the way for the massive transformation of the Poudre and South Platte watersheds, with Colorado's 1937 Water Conservancy District Act and United States Bureau of Reclamation funding for water projects begun thereafter. The Carey Act, in contrast, was entirely dependent on marketplace financing and bound to its statutory requirements, and it took off only after the 1893 depression lifted and the nation's irrigation bond market began its recovery after 1903. Most projects across the state were pure speculative promotions. But the four previously mentioned unfinished projects in northwestern and southwestern Colorado and the state's San Luis Valley helped keep the Carey Act relevant as a public policy, and they further illustrate the boundless minds of developers unwilling to accept nature's limits.[6]

Chapters 3 and 4 place the Carey Act's greater significance in Colorado at the Two Buttes and Muddy Creek developments, the historic homeland of Indigenous and later-settling Hispanic peoples. These chapters examine the critical nature of the law's regulatory, financial, and social dynamics across a historic region and show how chimerical thinking influenced the developers of each reclamation project. Unrealistic expectations and the Carey Act in Colorado go hand in hand. Developers, with assurances from hydraulic engineers, assumed that flash floods—which seemed regular in southeastern Colorado—would sufficiently irrigate their segregations from streams that often ran dry much of the year. As identified by historian James E. Sherow, this chimerical mindset drove progressive engineers who were developing the hydraulic-dependent regions of the West. Fred L. Harris, the lead developer at Two Buttes, nimbly tailored the scheme to fit the Carey Act's peculiarities, especially its oversight by the State Board of Land Commissioners of Colorado and the law's dependence on the hypothecated irrigation bond. Little appreciated has been Harris's sale of land parcels and water contracts to settlers—both men and women—in his attempt to create an exclusive community that was distinct racially and morally from the region's historic tradition. The resulting irrigation works cost roughly $500,000, and water first ran in its canals in 1912. Meanwhile, along Muddy Creek, the ephemeral tributary of the Purgatoire, a combination of unchecked imagination, the 1894 law's lack of corrective measures, and the hypothecated irrigation bond worked against each other until developers reimagined the scheme by using upfront financing. However, both projects always teetered precariously close to financial ruin. Each project lacked sufficient water, had too few successful farmers, carried too much burdening debt, and was characterized by constant tension between settlers and developers.[7]

The Carey Act, rather than extending reclamation development beyond established canals, proved to be impeding it—a pattern of failure across Colorado and much of the West that state and federal authorities well understood by 1914 but did little to address. As explained in chapters 5 and 6, the collapse of the irrigation bond market in 1911 and Colorado's successful petition for an additional 1 million federal acres at the behest of developers set into motion bitter protests from settlers living on undeveloped projects in northwestern Colorado as well as from other landowners nearby. In 1914, progressive-era reforms that restructured the State Board of Land Commissioners of Colorado as well as initial federal investigations that attributed the state's pattern of Carey Act failures to irresponsible promotors, insufficient financing, and incompetent engineering led the state agency to cancel roughly 25 percent of the acreage it had previously segregated. The Muddy Creek Project was among the canceled projects for its lack of progress. But such failures were the pattern across the American West. The region's governors, among them Wyoming's Joseph M. Carey, met with federal reclamation officials that year in Denver, but neither party showed interest in interfering with capitalism. Moreover, they outright rejected as socialist and un-American the advice of Elwood Mead, renowned reclamation expert and one of the architects of the Carey Act, who proposed direct federal aid to settlers based on his experience establishing Australia's closer communities. Thus, the patterns of failure continued.

Chapters 7 and 8 foreground the later stages of the Carey Act's long collapse in Colorado, which gave conservation-minded individuals and groups economic and environmental justifications for converting the land and water of private enterprises to publicly owned wildlife areas. The deviation to that endpoint took decades. The first deviation took place at the Muddy Creek Project, which developers reconstituted as a mutual irrigation company that was partially self-funded and built a $425,000 irrigation works as the first phase of their attempt to irrigate 24,000 acres. They received unprecedented assistance from the state land board before the project's cascading financial troubles rendered it functionally inoperable. The first-person account of settler Isabel Dodge O'Brien throws light on everyday life there and offers a dual glance at the positive qualities of pioneering that enriched her family's growth but also forced it to rethink the propriety of chancing a livelihood dependent on an ephemeral stream. The project's subsequent formation in 1928 as an irrigation district, though subject to the Carey Act, was done principally to secure federal funding, especially during the New Deal. The move proved futile as infuriated settlers effectively thwarted its further catastrophic circumstance amid the Dust Bowl conditions of the 1930s and acknowledged the error of attempting to reclaim the vast acreage by irrigation. At Two Buttes, in contrast, the divergence from reclamation to conservation proved more contemplative. Its founder and manager, Fred L. Harris, came to acknowledge the project's nature-dictated limitations early on, forcing the enterprise into involuntary bankruptcy in 1927 and downsizing it from 22,000 acres to 13,000 acres and, later, to roughly 3,000 acres. He then reimagined the reservoir's function during the Great Depression and Dust Bowl years as both an agricultural and a conservation hydraulic works. As shown by historian Mark Fiege, nature constantly thwarted such attempts by western farmers to transform their distinct irrigated units of capitalism, forcing them to act according to *its* design.[8]

Two Buttes and Muddy Creek Reservoirs were key examples of such units of capitalism, and they had always had a recreational feature in their early history. The stocking of fish for local anglers began immediately after the construction of each impoundment. Migratory waterfowl, following the North American Central Flyway, found the new reservoirs inviting for winter stopovers, and hunters occasionally shot them. However, the reservoirs' importance as wetlands conservation areas only occurred after 1930 with the alignment of world and national conservation values, public wildlife policies, the science of ecology, neighboring landowner buy-in, and sufficient funding capabilities to sustain a conservation purpose. The rise of contemporary environmentalism since the 1960s has further forced a reevaluation of these two relatively obscure locations as non-game and threatened species further define the relationship of humans to nature. Indeed, as historian Philip Garone examines in his environmental study of California's Central Valley, a vast area somewhat analogous to parts of the Great Plains, such shifting attitudes about the unrestrained manipulation of nature and the economic incentives to drain wetlands have been gradually overshadowed by ecological justifications to preserve them.[9]

Chapters 9 and 10 detail the gradual development of those lasting aesthetic values that deviated from the materialistic principles that created the Colorado Carey Act projects—a shift that led to their transformation into wildlife areas, special places where nature still regulates but also gifts its wonders. At Two Buttes, the project's failed financial condition and drastic paring down of irrigated acreage gave greater importance to its recreational use. Its manager, Fred L. Harris, a state legislator, successfully secured its designation as a state wildlife refuge. Such designations by the Colorado General Assembly (like the federal government's location of national parks and monuments) were nearly always given to its vertical landscapes of mountains and its great canyon country. Congressional enactment of federal funding for wildlife protection in 1937, 1950, and 1965, with its antecedents deep in the history of the American conservation movement and with critical help from local wildlife groups, provided a scientific as well as recreational rationale for Colorado to eventually purchase the area in 1970. This new approach to American wildlife management, as historian Jared Orsi has shown, contributed to the birth of ecological thinking—the ethic that considers nature and society to be interdependent and forms the core of modern environmentalism. Not to be undervalued in this ideological shift was the confluence of regional soil and water conservation efforts as well as the development of groundwater for irrigation. Meantime, at the site of the blighted Muddy Creek Reservoir, where no significant groundwater existed, state wildlife officials, under pressure from local conservationists, purchased and refitted the waterworks into the Setchfield State Wildlife Area in the late 1950s. However, the venture was short-lived, as nature reclaimed it after a killer flood in 1965 rendered its dam inoperable and repair too costly. Muddy Creek Reservoir's precious water right for wildlife became the all-important, critical asset for far southeastern Colorado's largest state refuge, the 19,000-acre John Martin Reservoir, as the region's expansion of wildlife habitat areas has come to exceed 144,000 acres. The contemporary Setchfield State Wildlife Area represents its own unique ecosystem, intermittent rivers and ephemeral streams (IRES), and illustrates the expanded meaning of these critical, worldwide environments, especially in Australia.[10]

The errors promulgated by the Carey Act in Colorado represent much of the content of this book. Missteps happen in life. The failure of the

Carey Act across the state is largely the story of a disconnect between private developers' hopes for windfall profits and the reality of unstable financing, economic rollercoasters, and the physiographic challenges of reclaiming high sagebrush lands. For most settlers who had staked their hopes and dreams on an agrarian way of life, the failure proved to be heartbreaking. Unfortunately, the historical record of settlers is thin, which is a shame.

And yet, remnants of the old Carey Act projects remain, their failures speaking to the missteps of humans and to nature's impermanence. Therein may reside a greater meaning of each to the broader history of development along the extensive watersheds that flow across the American West. The Arkansas River waterway—like the complex and vexing river systems of the mighty Colorado, the South Platte, the Rio Grande, the unruly Brazos and Pecos Rivers in Texas, and the lessening Santa Cruz in Arizona—is its own historical example of Americans' faith in their ability to manage nature. In the end, though, that ability must always yield in some fashion to nature's persistent cycles of flood and drought.[11]

Public and private conservation awareness along the waterways of the West, such as that in southeastern Colorado, demonstrates that a more sustainable use of wetlands now defines many places. In addition, the efforts of local southeastern Colorado residents and state and federal authorities to memorialize the historic sites of Boggsville, Bent's Fort, the Sand Creek Massacre, and the Japanese American internment camp of Amache further show a more thoughtful awareness of what such places mean. Environmental laws, private and public conservation organizations, and partnerships with landowners today balance aesthetic values and livelihoods in places where agricultural profit margins are thin. Perhaps this balance bodes well for the region's future, particularly for the benchlands and usually dry streams—places where boundless-thinking developers once believed it was possible to domesticate the landscape above the Purgatoire and on the wide shelf lands at the western edge of the Hugoton Embayment.

1

EXPERIMENT OR SWINDLE

Joseph M. Carey, a man who fought nature's limits, boarded a Union Pacific coach at Cheyenne, Wyoming, bound for Washington, DC, in early January 1894. At the time he was the state's only US senator. As the train crossed the Great Plains, or what early American explorers called the Great American Desert, he passed homesteads and towns—signs of an economic permanence that had already defined the irrigated river bottoms, bustling cities, and mining centers of the American West. But beyond those developments were vast arid stretches that were the public domain of the United States, seemingly irrigable areas of perhaps unlimited promise where settlers had yet to begin staking out their livelihoods.[1]

Across the immense prairie, the ages-old rhythms of life had forever changed. Society upon society of Native American tribes the United States had subjugated and stripped of their legal claims to the land had submitted to life on reservations. There, the government was attempting to force upon them assimilation policies, the unsuccessful attempt to extinguish the Indian identity. But almost entirely missing from the

https://doi.org/10.5876/9781646426492.c001

land were the great numbers of mammals common or indigenous to the prairie—bison, elk, deer, pronghorns, bighorn sheep, wolves, and grizzly bears. Once numbering in the tens of millions, many animals were on the brink of extinction, as beavers had been four decades before. The bison's tragic demise began in the 1830s with Native Americans bartering bison robes to white traders. It continued as infectious livestock diseases and depleted grasslands decimated the herds. During the 1860s and 1870s, market hunters finished off nearly all the remaining wooly animals.[2] For the senator, such profound displacements of people and animals were imperatives for the nation's expansion and capitalism's defining role in that development.

The rails over which Joseph M. Carey traveled east across the Great Plains were the same rails that had brought a growing market economy and him to the West in 1869 as President Ulysses S. Grant's appointed attorney general of Wyoming Territory. At that time, Carey was twenty-six, and he traveled west on the nation's first transcontinental railroad. As he passed through central Nebraska, he crossed the 100th meridian, the longitudinal divide that separates the humid and lush East from the arid West. The demarcation would be definitional to much of his life's journey, nearly all of which he would spend in arid Wyoming. He went on to serve as a justice of the Wyoming territorial supreme court, establish colossal cattle ranches from the public domain with his brother, speculate in irrigation schemes and town building, and serve as Wyoming's territorial delegate to the United States Congress. His life's purpose, beyond amassing a huge personal fortune, was to promote capitalism by advancing what he believed were the virtues of human intervention in nature to alter the meaning of arid western lands—of life as it had existed for Indigenous peoples for tens of thousands of years and for animals and plants for millions of years. Judge Carey, the title he preferred, was at constant war with the limits of the natural world. As he would later proclaim, "Man must conquer the desert, or the desert will conquer us."[3]

Arriving in the nation's capital, Senator Carey wore those deeply held materialistic beliefs about capitalism and exploiting nature for profit on his sleeve, values most western senators and representatives shared about developing the unsettled reaches of their states. Nevertheless, nature and economics constantly tested such beliefs. Carey's ranching empire, for example, had survived the devastating winter of 1886–1887, when an epic blizzard killed more than 75 percent of cattle on the northern plains and wiped out many fortunes.[4] And bearing down on the nation in 1894 was an unprecedented economic depression that had shut down nearly every sector of the economy. Many industrial millionaires were suddenly penniless, and unemployed workers sought out charities. Agricultural markets shrank. Recovery was nowhere in sight. In the upcoming session of Congress, the Republican senator from Wyoming meant to assist private enterprise's recovery, if perhaps only in a parochial way, by personally ushering legislation into law that would expand capitalism's dynamic onto the arid lands. During previous Congresses, two significantly broad proposals by western representatives to develop the arid lands had been gaining the legislature's attention: one, a law that the government cede all unsettled federal arid lands to the western states to develop themselves and offer to settlers; the other a law that authorized the government to sponsor widespread reclamation (irrigation) projects on the arid lands to induce settlers.[5] Whether Senator Carey sensed that neither proposal was likely to become law in the current session of Congress is uncertain, but he had brought with him a narrower proposal that permitted the federal government to gift 1 million acres of arid land to each western state to which the Desert Land Act applied.[6] That 1877 federal law, which Congress revised in 1891 (and made applicable to Colorado), defined arid land as land incapable of growing a crop without irrigation. It permitted individuals, not companies, to partake in reclaiming arid land but had proven wholly inadequate except for the smallest waterworks. Senator Carey's proposal, in contrast, favored a corporate role and provided that participating states oversee the private reclamation development on those lands for settlers.

The rationale that underpinned Senator Carey's 1-million-acre gift proposal, and indeed the two other contemporary reclamation ideas as well, was the American custom that created private property from the public domain. The policy of federal land cession has its origins in the nation's founding. The gifting began at the end of the Revolutionary War, with the various original states retaining large tracts of land. The Preemption Act of 1841, railroad land grants, and homestead legislation continued the policy. People and corporations quickly took advantage of the free real estate. The gifted land was a tangible commodity, an item useful or of value in the nation's market economy. And, of course, it was subject to the same exploitation and abuse of all products that conformed to laws of supply and demand. But this long history of federal gifting policy always had critics. By the early 1870s, opposition to cession among the nation's northern legislators had increased. Many of them attributed the nation's lack of a budget surplus to a lack of public land sales, which partly funded the federal government. Moreover, northerners had come to identify the gifting land policy as benefiting only western interests.[7]

Senator Carey knew well the contentious legislative issue of federal desert land cession to states from its origins in the early 1870s through the 1880s. When severe drought hit Colorado Territory in the early 1870s, the territory's governor, Samuel H. Elbert, emphasized that hundreds of optimistic farmers arriving on newly constructed railroad lines turned away from agriculture because drought conditions had forced them to return to states where preemptions and homestead laws had been mostly exhausted. In October 1873, irrigation interests gathered in Denver and petitioned the United States Congress to either grant states land or money or build irrigation works. Colorado's delegate to Congress introduced a bill to provide such assistance, but it died before reaching the United States House of Representatives floor. When another drought struck the West in the late 1880s, the issue of cession found broader congressional support. Congress entertained numerous policy considerations such as unconditional federal grants of all arid lands to states, direct federal involvement in reclamation, and the use of public entity irrigation bond financing districts similar to the one California had authorized under the Wright Act in 1887. Among the many unsuccessful bills was one Representative George G. Symes of Colorado introduced in 1888 that pleaded for 500,000 acres to aid the state in constructing storage reservoirs. Congress, nevertheless, did provide some short-term funding in 1888 for extensive western irrigation and reservoir field studies conducted under the direction of General John Wesley Powell, director of the United States Geological Survey. Two of the most formidable US senators from the West, Henry M. Teller of Colorado and William Stewart of Nevada, had promoted the survey. However, when the survey concluded, the idea of a comprehensive reclamation plan soon died when General Powell's long-held notions of ordered economic development of the region's river basins conflicted with prevailing

FIGURE 1.1. Joseph M. Carey, the godfather of the Carey Act, ca. 1890. Sub Neg 1528, photographer Walker. Wyoming State Archives, Cheyenne.

western attitudes (including those of the two senators) that judged such a policy as curbing private enterprise's rightful development of the West.[8]

Although the sentiment for cession was popular in Colorado during the late 1880s, Wyoming's political leaders kept the land-gifting idea before Congress. In the late 1880s, when Joseph M. Carey was Wyoming's territorial delegate to Congress and was preparing his first bill for the admission of Wyoming as a state, he included a section providing the new state with 2 million acres of land. He added the condition that the new state should see to the reclamation of the arid land and locate actual settlers there. A subsequent statehood bill eventually became law, but it omitted the arid land section. However, on July 10, 1890—the day President Benjamin Harrison signed the Wyoming statehood bill—Delegate Carey, soon-to-be United States Senator Carey, introduced another unsuccessful cession bill into the House of Representatives.[9]

Notwithstanding Senator Carey's early cession efforts, Francis E. Warren, US senator from Wyoming, also deserves mention in the arid-lands-gifting movement. Warren, a staunch Republican who had taken his seat in the Senate with Carey in 1890 when Wyoming achieved statehood, was briefly absent from the Senate from 1893 to 1895 when the state legislature did not reelect him. Warren's early role in arid-lands policy was to some extent foundational to the gifting proposal Carey brought to Washington, DC, in 1894. Warren, like many solons of the West, saw newly enacted federal laws—such as the conservation laws that created national parks and national forests, laws that abolished the Preemption Act and the Timber Culture Act, and modifications to the Homestead Act and the Desert Land Act—as direct impediments to each state's economic development. In the Second Session of the Fifty-First Congress (1890–1891), Warren introduced an unsuccessful cession bill that called for the unconditional gifting of federal lands to arid states. The following year, during the Fifty-Second Congress, he introduced a new land-cession bill that proposed dividing ceded land to states into irrigation districts and pledging to sell or conditionally sell all of that land to accomplish reclamation. The unsuccessful bill provided for settlers 160 acres of irrigated and 160 acres of non-irrigable land. After ten years, if settlers had not reclaimed the land, it might return to the federal government if the government agreed to irrigate it. Finally, the bill permitted a state to lease unlimited amounts of its gifted pastureland that adjoined the irrigated acreage.[10]

Senators Carey and Warren were at the forefront of congressional efforts to achieve federal land cession, but public support for the gifting policy grew dramatically during the early 1890s with the rise of western irrigation congresses—quasi-government gatherings of irrigation promoters and land developers as well as government officials. Initially leading this movement was William Ellsworth Smythe, whose national irrigation conventions and publication *Irrigation Age* heavily promoted hydraulic engineering and widely spread the gospel of reclamation's transformative potential. The movement espoused an orthodoxy, which held that developing the arid lands would expand the national economy and thus provide new homes and livelihoods for millions of white middle-class citizens, as Smythe would have it. Furthermore, it framed agriculture's growth as a moral issue. Farming, or so the long-held American myth maintained, was the essence of the American character, self-reliance, hard work, and patriotism. During the early years of this advocacy group, its leaders highly favored the policy of western land cession. Smythe and others believed the federal government would never appropriate funds to develop empty desert lands. But the notion of direct federal building of irrigation works was never entirely absent from these early policy forums. By 1896, when Smythe stepped out of the limelight, a consensus had begun to emerge that advocated a direct federal role in reclamation. Smythe's successor, George Maxwell, a California lawyer and proponent of nationally funded reclamation, took up the group's cause of arid land development.[11]

The National Irrigation Congress that Smythe and, later, Maxwell headed was a powerful lobbying organization. Indeed, after its founding in 1891, both houses of the United States Congress formed separate Committees on Irrigation and Reclamation of Arid Lands. Senator Warren's 1892 cession bill received the support of the new Senate committee, which he chaired. However, the bill failed to receive consideration because of its lower listing status on the Senate calendar. Nonetheless, not every western congressman favored Warren's bill. Senator Thomas C. Power of Montana saw corporate land monopolies as the beneficiaries of such a cession. He only needed to look at Wyoming to see cession's menacing reach. In Wyoming, where cattle and railroad monopolies controlled virtually every aspect of the state's economy and governance, small-time ranchers and farmers in Johnson County who objected were violently attacked by an army of hooligans working for the Wyoming Stock Growers Association (WSGA). Two of the WSGA's most prominent members were Senators Warren and Carey. The Johnson County War significantly fueled opposition to cession in Wyoming. In the November 1892 election, the state's few Populists joined Democrats to elect Democrat John E. Osborne to the governorship. Democrat Henry A. Coffeen was elected the congressional representative, and the legislature refused to reelect Senator Warren, a Republican, who was one of the loudest voices in Congress arguing in favor of the reclamation cause.[12]

Meanwhile, in neighboring Colorado, anti-monopoly sentiment rose most vocally around matters concerning the mining industry and silver's declining importance in the national economy. However, many Colorado farmers also embraced anti-monopoly thinking. In northeastern Colorado, numerous farmers objected to the powerful, large water companies—several of them foreign interests—that had come to control many of the principal ditches and large swaths of irrigable land. In 1892 significant numbers of farmers joined with the state's other anti-monopolists to help elect Populist Davis H. Waite governor. Waite deeply resented corporate interests yet fully supported increasing the state's population as long as settlers, not corporations, were the beneficiaries of such growth. But any grandiose plans he may have had for increasing migration to Colorado met the reality of the nation's most severe economic depression to date as mines, smelters, factories, and businesses of all kinds closed and demands for agricultural goods collapsed.[13]

The Depression of 1893 and the rise of Populism that led to Warren's defeat momentarily quieted congressional action on cession and its attendant element of reclamation. But Senator Carey plowed ahead nonetheless, certain of his cause. On February 1, 1894, nearly a month after he had boarded the coach in Cheyenne, he introduced Senate Bill 1544 in the Second Session of the Fifty-Third Congress. The bill was remarkably brief. It simply authorized the granting of lands to several western states and territories and "for other purposes." Senator Carey sat on both the Committee on Irrigation and Reclamation of Arid Lands and the powerful Committee on Public Lands. The presiding officer of the Senate assigned the bill to the Committee on Public Lands, one of the body's original standing committees. The committee, however, did not report out Carey's bill. Twelve days later he reintroduced the bill (S. 1591), which again went to the Committee on Public Lands. This bill quickly made its way out of the committee. The proposal offered 1 million acres of arid land for reclamation by each state in which the Desert Land Act applied (but it included Kansas and Nebraska), stipulated that "actual" settlers reclaim the land, and gave states, not the federal government, greater authority over its administration. On April 17, Senator Carey reported it back to the full Senate with a strongly favorable committee report backed up by a supportive letter from General Land Office (GLO) commissioner S. W. Lamoreux. The committee and the commissioner argued that given existing congressional objections to an excessive federal role in reclamation and concerns about direct federal expenditures for reclamation, the bill nevertheless addressed the needs of arid land settlers who needed assistance because of the steep cost of irrigation systems. Moreover, they reasoned that the legislation involved a small portion of each state's federal arid lands and that the experiment could not negatively affect the federal government because if a state did not see to the development, the acreage would revert to the federal government.[14]

Senator Carey's bill went on the calendar, and the full Senate debated and passed it without controversy on July 18. Colorado senators Henry M. Teller and Edward O. Walcott, whose state stood to possibly benefit from the law, did not address the legislation. However, the powerful chair of the Committee on Appropriations, Sen. Francis M. Cockrell of Missouri, had insisted on four changes: first, that a state have ten years, rather than three, to select the segregated land from the public domain; second, that it align with the cultivation requirements of the Desert Land Act and that a settler must spend at least three dollars per acre to improve it; third, that states have five years rather than three years to reclaim the land; and fourth, that the United States Department of the Interior, through the GLO, would "make all rules and regulations necessary to carry out" the law. The bill then went to the House of Representatives.[15]

Meanwhile, just weeks earlier, in late May, the House Committee on Irrigation and Reclamation of Arid Lands had taken up its own legislation that proposed much broader national irrigation measures than those in Senator Carey's bill. Committee member Representative Willis Sweet of Idaho, with the broad support of western members of Congress, introduced one substitute bill for three similar reclamation bills. His bill (H.R. 7558) called for a $325,000 appropriation to pay for land and river surveys and the preparation of plans for the construction of one irrigation project in each of the arid states and territories at specified locations cited in the bill. In Colorado, the survey was to occur on the South Platte River and its tributaries, a site probably chosen by Representative Lafe Pence of Denver, who sat on the committee. In late June, the committee recommended that the House pass the bill, but it never came to the floor. Meanwhile, in the Senate, at the end of July the staunch anti-monopolist Senator Power introduced nearly identical legislation (S. 2248) into the upper chamber to advance Congressman Sweet's legislation. The bill went to the Senate Committee on Irrigation and Reclamation of Arid Lands, which reported it back. But the Senate did not take up the bill, perhaps because of its late filing or the fact that senators considered the agreed-upon Carey bill a compromise to the more sweeping Sweet-Power bill.[16]

In any event, by August 2, 1894, the House Committee on Irrigation and Reclamation of Arid Lands still had not reported back on Senator Carey's bill. The business of the Congress was to end in a few days. Senator Carey, discouraged by the House committee's inaction but believing it had nonetheless favored his bill, decided to seek its attachment to the sundry civil appropriation bill (H.R. 5575) then under consideration in the Senate. On the floor of the upper chamber, he asked that his bill (S. 1591) be added as Section 4 of the House-originating bill that funded the entire federal government. His amendment was urgent, he explained, to permit state legislatures, which met during the winter, to enact necessary legislation to administer the program immediately. To justify its attachment to the appropriation bill, he included a nominal appropriation of $1,000 for a survey of land by the United States Department of the Interior. Apparently, the members of the Senate Committee on Appropriations had initially declined the amendment's inclusion on grounds that it was out of place in a general funding bill. Nevertheless, they agreed to the amendment, provided it was sufficiently brief. Without objection or a roll call, the Senate agreed to Senator Carey's amendment. Along with the general funding bill, it went to the joint conference committee of the House and the Senate, where late-arriving appropriation matters always moved quickly.[17]

In the conference committee, the house managers also disapproved of Senator Carey's amendment, which the Senate had attached to the appropriation bill. Nonetheless, they crafted a substitute amendment that they backed up with a favorable letter from Edwin A. Bowers, the newly appointed acting commissioner of the GLO. The amendment gave the Department of the Interior greater authority over state projects with a regulatory clause that gave midwestern representatives a greater say in the legislation. As the Desert Land Act stipulated, the substitute amendment required states to supply the Department of the Interior with maps that located the proposed irrigated land, showed the mode of irrigation, and identified the source of irrigation water. But it did not include the requirement of three dollars per acre to go toward reclamation. Moreover, the amendment excluded Kansas and Nebraska. The conference committee immediately sent the amendment back to the full House, recommending its approval so the chamber could soon vote on the appropriation bill.[18]

The House began debating the amendment on August 10, and the representatives immediately struck out the Senate version and took up the House conferees' substitution that required a greater GLO role in the law's administration. The debate was vigorous and followed different arguments. For many members of Congress, such as Illinois Representative Joseph G. Cannon, the previous chair of the Committee on Appropriations, the amendment was simply out of place attached to an appropriation bill. He thought it was "improvident" to undertake granting 1 million acres to states after an hour or two's discussion. He believed that settling the arid lands was important but saw little prospect of their development until the nation's population of 70 million exceeded 140 million. Cannon did not necessarily oppose giving the land away, but the proposal should stand on its own merits as a standalone bill. He feared that such hastily enacted legislation would hang around the necks of future generations in the arid lands "like a millstone." For other members of Congress, such as Thomas C. McRae of Arkansas, chair of the House Committee on Public Lands, the proposed law was an experiment: "The object is to encourage the irrigation of these lands of the States." He explained that unlike the Senate provision, the House substitute authorized the secretary of the interior, with the approval of the president, to make a contract with a state for reclamation not exceeding 1 million acres in that state: "No reservation or withdrawal is permitted except for temporary purposes until the plan is approved, and no title [land patent] is to pass until thorough irrigation is accomplished and shown to the Interior Department. . . . We want settlers upon all of our public lands. This is simply an experiment by which it is proposed to aid the States in their efforts to reclaim the land, by remitting the price [of land] and cost of survey." But some representatives argued that the new law would be a swindle, a "bunco game" said Iowa's William Peters Hepburn. The amendment's fiercest critic, Representative Henry A. Coffeen of Wyoming—whom the Democrat Populists helped elect while ousting Senator Francis E. Warren—was absent that day because of illness.[19]

Notwithstanding the amendment's critics, members of the House who were from the West overwhelmingly threw their support behind the substitute version despite its attachment to the appropriations bill. Colorado's John Calhoun Bell, who represented the state's Western Slope, expressed their sentiments best. Although he was supportive of the broader Willis Sweet bill, he may have had concerns that it excluded his region. But Congressman Bell took the floor and enthusiastically supported the amendment. A noted lawyer who had settled briefly in the San Luis Valley and then in Lake City before locating to Montrose in southwestern Colorado in 1885, Bell reminded his colleagues of irrigation's fundamental importance to the region's economy and thus to the nation's development: "Along every stream of the irrigating region, the ground which an individual can irrigate is taken now; but there are millions of acres as bare as this floor, where by diverting water at great expense and spreading over it, [they] can be made very fertile, productive, and profitable." Today, he said, the roads are lined with men from western Nebraska, western Kansas, and elsewhere looking for ground that can be irrigated. "In the great valleys of Colorado," he continued, "if you give the states these lands, the Government of the Unites States will lose nothing." He singled out the millions of dollars spent by Travelers Insurance Company on irrigation projects in the San Luis Valley and near his ranches in the vicinity of the Uncompahgre River: "The ditch from which my immediate neighbors water their lands cost $200,000. They [irrigation companies] furnish the farmer water for $1.10 an acre per year." He added that "no individual can build these canals; some corporation, public or private, must [do so]." The lands, he said, were a barren waste until farmers artificially watered them, and the nation should devise a method to reclaim them: "We know this plan or any other national one will have to have years in which to mature. The present plan is but an experiment—a reminder . . . that the great agricultural production of the future in the United States will be, must be, through irrigation and the subject must be dealt with as a national question." Bell could see the future. But he could not have foreseen at that moment when, a decade later, he would lead the congressional effort to authorize construction of the massive Uncompahgre Project near Montrose by the newly established United States Reclamation Service.[20]

When the long debate finally ended on August 11, the House approved the substitution provision of the amendment 159 to 9 and referred it back to the conference committee one last time. In turn, the conferees of both chambers agreed to the House version, adding only the clause that not less than 20 acres of every 160-acre tract be cultivated by actual settlers. For the last time, the committee sent the amendment back the House for inclusion in the appropriation bill. Except for relatively minor additions, Senator Carey's original and fairly simple bill had finally taken form.[21]

Four days later, on August 15, the House of Representatives considered the appropriation bill in its final form. Its approval was a certainty. In the chamber that day was Wyoming's Henry A. Coffeen, fully recovered from his illness. He took the house floor and with pent-up anger unleashed a blistering attack on the measure, which he rightfully saw as being railroaded through the chamber. The congressman was not a member of the Public Lands or Irrigation and Reclamation of Arid Lands Committees, although he had strongly favored Representative Willis Sweet's unsuccessful proposal of direct federal involvement in reclamation. He had convinced the Idaho representative to specify Wyoming's North Platte Basin for that bill's proposed land and water survey. But this was his day to vent. Coffeen was a merchant who had settled in Sheridan, Wyoming, in 1884 from Danville, Illinois; been an official with the Knights of Labor; and once run for an Illinois congressional seat. Like many other northern Wyoming residents, he had deep sympathy for those who had thwarted the invaders hired by the Wyoming Stock Growers Association during the Johnson County War. Coffeen was an enthusiastic advocate for irrigated agriculture in desert lands and considered the House version of the Carey legislation that gave a greater role to the GLO to be more protective of the settler. But still, he thought it was an outright swindle by land-hungry syndicates, such as the operations of cattle empires in Wyoming and across the West: "Where and when have public lands ever been turned over to the States in hurried and loose methods without resulting in gigantic land frauds and plunder by land grabbers?"[22]

Coffeen correctly foresaw the settler-cultivator as perhaps never gaining patent to the land, given that people who invested capital for reclamation in advance would certainly secure their investment with liens on the land. The merciless greed of corporations, he insisted, would reduce the entire country to tenantry if given full sweep in such legislation. So, what should be done with the arid lands, he asked? "I would have the National Government enter upon the reclamation of the arid lands of the West." Settlers from all parts of the United States would surely come, "so let the United States expend the money necessary," for the "General Government is abundantly able to do so while the States are not." Unfolding a large map of the West's major river basins, he pointed to irrigated agriculture's great significance for the many states and territories. He saw great possibilities for the nation with irrigation's potential expansion to arid lands. To be sure, unused water existed, but the new legal concept of water-use rights known as appropriation, he pointed out, presented fundamental questions about development that deserved thoughtful debate. Sounding like John Wesley Powell, he reminded those who listened that nature had no reverence for state boundaries. The distribution of water, he said, was also an interstate affair, and it was the national government's rightful place to control that distribution: "Let the people own the land, but water must be owned by the general masses." Coffeen spoke for nearly an hour before concluding: "I shall vote against the adoption of the conference report . . . I shall work and vote against all measures that would in any manner permit land-holding corporations to come between the people and the ownership of land so necessary to their welfare and the safety of our Republic."[23]

Thus, a frustrated Coffeen finally ended his objection, and the House of Representatives approved the sundry appropriation bill. The Senate quickly followed. On Saturday, August 18, 1894, President Grover Cleveland signed into law H.R. 5575 that included Section 4, thereafter informally known as the Carey Act.[24]

The Carey Act—however messy and awkward its journey into law—nonetheless offered palatable legislation to various constituencies. For eastern members of Congress, it meant no costly outlays for a massive public works project in the West. For proponents of cession, the act granted at least some acreage. For states' rights advocates, it gave states general control of the proposals. And for those who believed reclamation should be a function of the federal government, it at least left it for the GLO to agree to a project's development. Moreover, the law did not preclude a future federal role in directly developing great swaths of the arid lands.

But beyond the Carey Act's legislative offering to the various constituencies, it was a notably exclusive law. People who might partake in its gifting of essentially free land were ironically those with financial means: land developers adept at marketplace transactions, financial agents with access to capital markets, experts in hydraulic engineering, and middle-class settlers with sufficient funds to buy into a project and build farms. Like the solons who crafted the law and the participants who would put it into practice, the "actual" settlers, as the law referred to them, would always prove to be white. The public policy was less explicitly racist than the nation's Reclamation Act of 1902, which, for example, prohibited Asians, but its exclusive construction expressed the ethnocentrism that Senator Joseph M. Carey—and indeed most power brokers in the West—promulgated.

The class and racial exclusivity of the reclamation movement was always on display at the National Irrigation Congresses. In early September 1894, three weeks after enactment of the Carey Act, the Third National Irrigation Congress met in Denver. The economic depression had hit the Mile High City with full force. But some people may have gained optimism over the buzz with irrigation talk as developers, engineers, politicians, irrigation equipment salesmen, and other delegates filled downtown hotels. Heading the congress was William Ellsworth Smythe, who just months earlier had been one of the most outspoken opponents of the Carey Act. The self-appointed publicist of the irrigation movement and devoutly racist advocate for the white middle class strongly favored the general cession of federal arid lands for reclamation, but his attitude toward the Carey Act was becoming more favorable. By November he would be calling it the most important opportunity for progress and "providential" to middle-class settlement of the arid lands, thus asserting the superiority of Anglo-Saxon values of material progress over Native American and Hispanic peoples who for generations had successfully adapted their lifestyles to the desert's limits. Notwithstanding the gathering's exclusive class and racial nature, an ideological divide over cession still lingered, and passage of the

Carey Act, rather than the Willis Sweet bill, sent tempers flaring. Amid the throng at one session on September 5 was Representative Coffeen, who reiterated his argument in the House that the Carey Act would effectively place the arid lands into the hands of syndicates. From nearby came the booming voice of his foe, Senator Joseph M. Carey. To an applauding crowd, Senator Carey explained that the federal government was unwilling and unable to do the work necessary to reclaim the arid lands. Besides, the people could do a better job of it. The depression would pass, and development would occur. He implored his opponents to give the new law a fair trial before they condemned it: "Man must conquer the desert or the desert will conquer us."[25]

Senator Carey and Representative Coffeen did not come to blows. But in addition to Coffeen's criticism of the Carey Act as public policy, he knew that the senator personally stood to reap financial gain from its passage. Since the early 1880s, Carey had had a stake in the Wyoming Development Company, a scheme Wyomingites knew well, that proposed taking water from the Laramie River to irrigate more than 50,000 acres in the southeastern part of the state. Although the company had dug about 100 miles of canals, it could not secure title to the nearby public lands. The company's story, as historian Donald J. Pisani has written, "illustrated the need to unify control over land and water so that private companies could use large blocks of public land as collateral for their investment. Carey's land bill was tailored to meet this need." Wheatland Colony settlers began arriving in 1894, and it quickly became a Carey Act project.[26]

Thus, the Carey Act, from its conception to its very beginnings as law, was rife with controversy. Two months after the irrigation congress had adjourned, Colorado voters elected Governor Albert W. McIntyre, a cattleman and Republican from the San Luis Valley who had no objections to the new law. But the law's controversy continued. Ousted Governor Waite blasted the legislation, reiterating his Populist critique: "In my judgment the Carey land bill is in the interest of private corporations who desire to gobble up the arid lands for cattle ranges and speculative purposes. There can be no objection to the use of the public lands for cattle range, even free of cost, if such does not prevent or hinder the occupancy by the actual settler. . . . No private corporation ought to be allowed to forestall his [settler's] right of settlement." He recommended that the legislature accept the land only on the condition that the state heavily regulate the law by retaining the land and leasing it to settlers. But, like Henry A. Coffeen, the outgoing governor could influence no one; they were two Cassandras warning of trouble to come.[27]

Meanwhile, most Coloradans interested in reclamation followed news about the Carey Act in newspapers. Colorado ratified its acceptance of 1 million acres of land on March 15, 1895, when John R. Gordon, a state senator from Pueblo, introduced the measure and it went on the books without opposition. Those mindful of irrigation and development matters soon learned much about the law's application in the state when two projects, one on the South Platte and one on the Arkansas, received general coverage. The details of those proposals, as the reader will see in chapter 2, were grandiose. But people also read about Carey Act projects in other arid states, most specifically the act's difficult beginnings in Wyoming where developed irrigation was in its infancy and watering tertiary land had proved difficult.[28]

Wyoming ratified the Carey Act in February 1895, one month before Colorado did so. Editors in the Centennial State noted the zeal of Wyomingites who backed the law even before that state's legislature approved the land grant. In Leadville, Colorado, where silver mines and smelters remained closed as the depression worsened, the *Leadville Evening Chronicle* announced "Wyoming's Future Prosperity" and proclaimed that the act would attract investment capital to the state without Wyoming having to assume obligations: "Settlers will flock to Wyoming, its waste water and unused land will be brought together, and from there wealth will flow into the coffers of the state."[29]

In June 1895, Wyoming approved the nation's first Carey Act proposal—Salon L. Wiley's Big Horn Basin Development Company Project. The state permitted a diversion of the Grey Bull River in northern Wyoming to water irrigable lands on the Germania Bench. Wyoming's implementation of the law, like that in most other Carey Act states, authorized its administration by a land board and required a project's approval by the state engineer. Wyoming's state engineer was Elwood Mead, who had written most of the water codes in Wyoming's constitution in 1889. He was also the technician who drafted the unsuccessful cession bills Joseph M. Carey introduced when he was a delegate to the United States

Congress and those of Senator Warren from 1890 through 1892. Mead had been presiding officer of the 1894 National Irrigation Congress in Denver and was enthusiastic about the Carey Act. He, in consultation with Cheyenne attorney J. A. Van Orsdel, drafted Wyoming's statute ratifying the 1-million-acre federal donation. Other states, including Colorado, copied Mead's design verbatim. Prior to his Wyoming tenure, which began in 1888, Mead had worked as a surveyor and a teacher before earning a BS degree in agriculture at Purdue University in 1882. After serving several months in the United States Army Corps of Engineers, he began teaching a mathematics course at the State Agricultural College of Colorado in Fort Collins and became an assistant to Colorado's state engineer, Edwin S. Nettleton. In the meantime, he returned to the Midwest and studied for a degree in civil engineering at Iowa State College, then earned a master's degree from Purdue. Mead returned to Colorado in 1885, where he resumed his duties as assistant state engineer and at the State Agricultural College, where he became a professor of irrigation engineering—the first such appointment at an American university.[30]

For Elwood Mead, reclamation was more than simply engineering new structures and launching new water projects. It concerned solving existing problems and improving functioning projects through the marriage of *irrigation* and *settlement on arid lands*. The agrarian utopianism he embraced and promoted was a value he acquired during his boyhood on the family tobacco farm in southeastern Indiana. There, he witnessed the destructive consequences of exploitive agriculture that depleted soil, depended on tenant labor, and disintegrated cherished rural communities. At age twenty-two he moved to irrigation-dependent Larimer County, Colorado, and found his life's work, a calling that would enable him to devote all his energies to the betterment of rural life for family farmers. His dual occupations as teacher and assistant state engineer complemented each other perfectly. Although the Fort Collins school hired him primarily to teach mathematics rather than engineering, his creative competence and successive graduate work quickly led to his appointment to teach courses in irrigation engineering, an idea he had proposed to the school's governing board. Hence, his simultaneous work for the Office of Colorado State Engineer thrust the young engineer into becoming the blossoming proponent of irrigated agriculture. Mead learned much under the direction of State Engineer Nettleton, who had joined the utopian Union Colony that founded Greeley in 1870, been the colony's engineer, and designed its vast network of self-financed and communally built ditches. Mead measured water flows of the colony's many ditches, but he also mapped other northern Colorado irrigation systems—some of which were financed by English capital. He was often tramping through the mountain streams that fed the region's canals. Thus, Mead obtained and integrated into his teaching practical experience that complemented his theoretical understanding of reclamation.[31]

By the time Mead moved to Cheyenne, Wyoming, in 1888, he had come to understand reclamation as offering the promise of agrarian virtues held by farmers who had already reclaimed what he estimated were 32 million acres of valueless and unproductive land in the West. He claimed that with the art of irrigation, perhaps another 250 million acres were reclaimable. For Mead, the term *settler* implied *virtuous* middle-class settler, and his attitudes about race reflected the mainstream racial prejudices of the time.[32]

Mead was both engineer and promoter, a dual role common at that time. Weeks after he approved the Big Horn Basin Carey Act Project, he and other top Wyoming officers and developers faced accusations of fraud in connection with managing the Carey Act. In Wyoming, a principal leader of the anti-cession sentiment was state senator Robert Foote. A Buffalo, Wyoming, merchant, Foote had helped whip up the frenzy of small-time ranchers and farmers in northern Wyoming to confront invading gunmen hired by the Wyoming Stock Growers Association during the 1892 Johnson County War. Foote leveled his charges against Mead in a detailed and widely circulated August 1895 letter to Hoke Smith, secretary of the United States Department of the Interior. Foote called the Carey Act a gigantic fraud that was simply a front for State Engineer Mead—an absolute dictator—to deprive actual settlers of the public domain.[33]

In September 1895, amid the furor over charges of fraud, Mead approved Wyoming's second Carey Act project. It was a proposal also in the Big Horn Basin that sought to divert water from the South Fork of the

FIGURE 1.2. Elwood Mead, ca. 1890. Sub Neg 2463, photographer unknown, Wyoming State Archives, Cheyenne.

Stinking Water (renamed Shoshone) and to irrigate lands segregated from the public domain. The Shoshone Land and Irrigation Company was an interest of William F. (Buffalo Bill) Cody, with George T. Beck as construction superintendent. Cody financed the project with money from his Wild West Show. The participants' and project's drama made for good headlines in newspapers across the country.[34]

Foote's dramatic accusations, which Mead and other officials rebutted, were not without general merit. His argument underscored the assumption by some farmers and ranchers that farmers themselves might build irrigation projects in isolated areas if water were truly available. In 1896 the editor of the *Greeley Tribune* voiced that very point about the tertiary lands in Colorado. The editor's position was that Greeley-area farmers had self-financed irrigation projects entirely without gifted federal land or federal financial aid. But the immediate consequence of Foote's inflammatory letter was to bring to a near halt the processing of all Carey Act applications by the Department of the Interior (through the GLO), which required that companies file extensive maps, requests for rights-of-way, construction plans, and other administrative requirements. After considerable project revisions and heavy lobbying efforts by the Cody Canal Company, President Grover Cleveland signed the contract with the company in March 1896. The famous showman's project was the first Carey Act development in the nation. Other Wyoming projects, including the Big Horn Basin scheme, momentarily stalled, and Mead became convinced that the GLO's regulations effectively undermined the act. Wyoming received no other applications until 1898, and by then Mead had returned to his belief in the absolute cession of arid lands to the states. Wyoming contracted with dozens of companies and eventually segregated a total of 1.39 million acres, though it issued settlers patents to only 203,311 acres. The most prosperous of the colonies was the Wheatland development. But the company itself, which had once had Senator Carey's interest, was not profitable. Its losses up to 1951 were $1.6 million.[35]

The lack of significant early progress in Carey Act development in Wyoming and the law's dormancy elsewhere was attributable to the protracted depression and decimated capital credit markets. But the law had weaknesses, and Carey Act developers would spend decades going to extraordinary lengths to adjust the law in an effort to advance their schemes. The first amendment they pressured lawmakers to address was that the Carey Act make provision for the creation of a lien on the segregated lands. A lien, as Representative Henry A. Coffeen had noted, would be necessary as security for borrowed money to construct a project. Lenders were the financial stakeholders in a project and insisted on collateral before they purchased construction financing bonds. But such a lien was impossible, since settlers did not have patent to their land prior to actual operation of the waterworks, settlement, and cultivation of that land. Moreover, the act limited the time of accomplishment of reclamation to ten years from the law's passage in 1894. Such limitations, asserted proponents of the act, made it impossible to finance construction of such large projects. Congress, after hearing from representatives from Washington, Wyoming, and Idaho who had been under heavy pressure from developers, amended the act in June 1896 to allow all participating Carey Act states to create a first lien on the lands to be irrigated. The legislation, however, was meaningless because no state would collateralize a project. Moreover, the essential problem remained. No state could provide for making the cost of a waterworks a direct lien on the land because a lien could not be made under the law until water was available to "actual settlers." At this time, Congress also permitted states to create proof of the completion of a waterworks and to apply for land patents without regard to settlement or cultivation. This amendment was also ineffective, since developers needed a lien before construction, not after. Nevertheless, Carey Act developers repeatedly lobbied their state legislatures and the United States Congress. In 1901, Congress provided one meaningful change to the law for developers that permitted a reclamation project to extend completion to ten years from the date the secretary of the interior approved a state's application (rather than from the time a company made application) for segregation, and it provided an additional five years for a project's completion at the discretion of the secretary.[36]

After the 1901 amendment and with the depression finally lifting, the number of proposed Carey Act projects increased dramatically as developers with boundless minds set out across the West. Promoters with eyes to an improving irrigation bond market located most of the early schemes in Wyoming but also in several other states. The promotion of these projects received heavy coverage in regional newspapers. Colorado publications reported frequently on the southern Idaho scheme known as the American Falls Project, a plan to irrigate tens of thousands of federal acres from the Snake River. Also garnering much of the limelight were stories about schemes in Utah and Montana. Such reports were always promotional notifications of the act's increased utilization, never the reality of its limited success.[37]

Developers in Colorado were slow to embrace the Carey Act. The easily irrigated river basins that farmers had not yet developed were quickly becoming developed. Water, once easily obtainable, was becoming scarcer. And yet there existed millions of acres of state and federal benchlands and desert lands—an irresistible temptation for eager developers. In the lower valleys of the South Platte and the Arkansas, the state's first Carey Act proposals in 1895 and 1900 rose as grand promotions, only to collapse as promotions. There, the errant steps began. The pattern of failure would repeat itself across the state as irrepressible developers sought assorted ways to utilize the act and cling to unviable or unsalvageable project after project.

2

CAREY ACT BEGINNINGS IN THE CENTENNIAL STATE

Daniel A. Camfield and George H. West, two of Greeley, Colorado's most prominent figures and longtime farmers and ranchers in the South Platte River Valley, were the first developers in the state to file a reclamation proposal under the Carey Act in September 1895, six months after the state had ratified the law. The men were well acquainted with the river's flow and its many tributaries in northeastern Colorado. Moreover, their obsession to transform arid lands distant from the river would be the irrepressible trait of dozens of others across the state seeking riches under the 1894 law. Although other developers would not begin to file Carey Act proposals in Colorado until after 1900, this fixation on windfall profits from turning arid land into an irrigated Eden would be the point source of a later value conflict between settlers and developers that defined and helped doom the Carey Act in Colorado.

Camfield and West were less well-known than Colorado's most celebrated booster-irrigators: William E. Pabor, a Union Colony founder of Greeley who had moved to western Colorado; William Newton Byers,

https://doi.org/10.5876/9781646426492.c002

booster extraordinaire and founder of the *Rocky Mountain News*; and Theodore C. Henry, designer of irrigation schemes across the state. Nevertheless, the two successful businessmen, with financial interests in banking and cattle ranching, filed their Carey Act petition with the state even though officials were not ready to adequately administer the law. They submitted a Carey Act application and hastily prepared maps and plans with the State Board of Land Commissioners of Colorado as well as the state engineer to develop arid benchlands adjacent to the South Platte River. The men headed the Pawnee Pass Irrigation Company and intended to reclaim vacant state and federal grazing lands in Morgan and Logan Counties northeast of Greeley. The scheme proposed the segregation of an astounding 300,000 acres from the public domain, almost one-third of the total 1894 grant to Colorado. They estimated that the cost of the system would be roughly $2 million. Their plan was to sell water rights to middle-class settlers at twenty dollars per acre on essentially free land and pocket the difference between the sale amount and the construction costs. The settlers, in turn, would own their farms and the waterworks outright, thus spinning out regional economic vitality. The notion of expanding irrigation beyond the region's hundreds of existing canals and ditches had been incubating in both men's minds for years. Indeed, the unsuccessful Willis Sweet bill that the United States Congress had entertained during debate on the Carey Act included this vast area in its design for a national reclamation law. To what extent Camfield and West influenced that selection is uncertain, but they imagined that their design under the Carey Act would yield significant windfall profits to themselves and other investors by renewing the 1880s irrigation boom that had shaped much of the agricultural region and contributed to the state's economic prosperity.[1]

The ambitious Camfield and West were the products of irrigation's birth and rapid expansion in Colorado that coincided with the mining boom. Free gold and silver had lured miners, and free water had lured farmers. The marketing of irrigated crops to the mining industry began earnestly with the collective efforts of farmers constructing ditches along bottomlands in the 1870s, such as Greeley's Union Colony. It expanded to supplying cities as corporate irrigation developments built enormous canals hundreds of miles long toward hundreds of thousands of acres of corporate-owned land. The corporate era, as historians label it, was a period of extraordinary capital inflow, great sums of which were foreign. For example, between 1879 and 1883, the British-owned Colorado Mortgage and Investment Company and its subsidiary companies constructed three large canals in the South Platte Valley: the Larimer and Weld Canal on the Poudre River, the High Line Canal on the South Platte River, and the Loveland and Greeley Canal on the Big Thompson River. Such companies sought profits from the sale of land purchased from railroads and charged a perpetual fee to farmers for the use of water. This construction began the period of remarkable speculative boom in the valleys of the South Platte, the Arkansas, and the Rio Grande, as well as on the Western Slope, that involved both foreign and domestic capital. Companies secured water rights, which were free, but the rights were junior in time to senior users who had exclusive use of water in times of drought under Colorado's Doctrine of Prior Appropriation System. Corporations, having promised more water than they could deliver and with canals they found difficult to maintain, proved financially unfeasible. The speculative corporate play had aided the rise of the Populist Party, as mentioned in chapter 1, and by the late 1890s most of the corporations had failed. Their water users then generally assumed the assets of the insolvent companies after receivership proceedings, and the farmers formed mutually owned irrigation companies—the same entities Carey Act settlers might form after paying off their water-right debt to developers.[2]

Camfield and West saw in the Carey Act a marketplace capitalism that might spur private enterprise once again, even though the national depression was crushing livelihoods across Colorado. They believed the growth of intensive irrigated industrial agriculture was inevitable and would continue to transform the entire South Platte River Valley. The driving force of that inevitability, they reasoned, was the impending growth of the sugar beet industry. Reclamation through greater water storage capacity offered by reservoirs was central to turning the region into splendid farming country filled with sugar factories. Since the 1870s, water users in the valley had tapped the greater portion of the South Platte's capacity to deliver water during irrigation season. Those earliest water users, having senior priority, had consumed much of the river's flow from its snowy mountain origins and that of its tributaries; thus, Camfield and

FIGURE 2.1. "Let no fertile land go dry," a favorite phrase of Daniel A. Camfield. From Watrous, *History of Larimer County, Colorado*, 347, photographer unknown.

West's claim to water was very junior in standing. However, by the early 1880s, developers had begun to foresee the dependability of increasing return flow to the river from upstream irrigation and founded the canal-dependent South Platte towns of Fort Morgan and Brush east of Greeley. Camfield and West thought that even more winter return flow was possible, an assumption they believed would supplement off-season flood flows—including from the intermittent Wild Cat and Pawnee Creeks—to provide new settlers with sufficient water capacity. In the case of the nearly free Carey Act lands, the two men reasoned that the cost of farming would be more a function of the initial cost of water, which their company meant to sell and which they claimed existed in abundant supply.[3]

Water, like cheap land, was a lucrative commodity for entrepreneurs such as Camfield and West, who obsessed about locating settlers on arid lands, notwithstanding that their use rights were junior in appropriation. The fixation with materialism was inseparable from their belief that they were advancing a common national good. But the belief concealed a bias

FIGURE 2.2. George H. West, ca. 1895, saw the fusion of horticulture and town building as vital to luring people to the unsettled reaches of Colorado. AI 3935, City of Greeley Museum, Permanent Collection, photographer unknown.

that neither people nor others with the same mind-set would bring themselves to acknowledge or appreciate. People had been living in the region for at least 12,000 years without radically exploiting the environment for the sake of profit and limitless development. First was the megafauna-hunting Clovis culture, of which archaeologists found glimpses with mammoth remains at the Dent Site; the similar nomadic Folsom people at Lindenmier; the Neolithic people, whose hunting, gathering, and agriculture lives populated the Great Plains; and then the Apache and later the Plains tribes, whose lives and cosmology intertwined with those of the bison. For the Cheyenne and Arapaho in Colorado, life across the region's once great commons ended with the tragic Battle of Summit Springs in 1869 south of Sterling, adjacent to land that would become Camfield and West's Carey Act proposed acquisition.[4]

The historic and sacred homeland of the Cheyenne and the Arapaho meant as little to Camfield and West as the dramatic environmental changes to the South Platte River itself since the time of white settlement. Water diversion for irrigation had channelized much of the once broad, shallow flow of the river, which in some places had been hundreds of feet wide. Gone were its numerous shifting sandbars and the bipolar flood-dry nature of erratic flow that allowed for a diversity of plant and animal life along the river. The changes also impacted native fishes and flocks of tens of thousands of migratory sandhill cranes, geese, and other birds that had once settled briefly along the river twice each year to rest and feed. The river's human-caused metamorphosis—like that of the Arkansas, which originated in snowcapped mountains and spilled onto the Great Plains—was a common consequence of irrigation's imprint.[5]

Camfield and West gave no thought to the South Platte's natural history. In fact, they meant to alter it even more. However, their Carey Act application and plans lay dormant at the office of the State Board of Land Commissioners of Colorado from 1895 to 1897. A combination of factors stymied the approval process. The agency lacked the necessary resources from the state legislature to fully implement the law, the General Land Office (GLO) in the United States Department of the Interior had begun its moratorium on all Carey Act projects because of the Foote-Mead controversy (mentioned in chapter 1), and the company redesigned its $2 million system to include Sanborn Draw Reservoir (later Riverside) and Jackson Reservoir at an additional cost of $300,000 to supplement its downstream Pawnee Pass impoundment. In the meantime, Camfield and West renamed their enterprise the Platte River Land, Reservoir and Irrigation Company and hired nationally famous irrigation expert and promoter Colonel Richard J. Hinton as its chief financial officer and a possible link to deep-pocket eastern syndicates. Hinton boasted that the three off-channel reservoirs could prepare land for the comfortable settlement of as many as 7,000 families.[6]

While the colonel boasted, Camfield and West were constantly recalibrating and reinventing their scheme, adjusting their earlier misjudgments about reclaimable acreage, rethinking investor and settler disinterest in the idea during the depression, and anticipating a noticeable shift in congressional consensus on direct federal reclamation of arid lands. Meanwhile, in early 1897, the state land board finally granted them permission to proceed but only on 100,000 acres the state engineer believed were reclaimable. The GLO approved withdrawing the Pawnee

Reservoir Site from the public domain but refused the state's request to segregate the 100,000 acres it had permitted to the company. Apparently, the GLO doubted the project's feasibility and refused to execute a Carey Act contract gifting the acreage to the state to give to the company. Nevertheless, in addition to the company's inability to advance the project, the cession of federal arid lands to western states remained a hot-button issue among developers and western politicians. Colorado Representative John F. Shafroth, for example, introduced an unsuccessful bill to grant the state an additional 1 million acres, as well as introducing numerous unsuccessful general cession bills between 1897 and 1898. Perhaps Camfield and West believed this outright gifting of public domain would become a reality as a later United States Geological Survey (USGS) study in 1900 found that their string of proposed reservoirs might be feasible and would be an important economic component of the rapidly expanding sugar beet industry. In any event, in 1902 Camfield and West suddenly withdrew their stalled Carey Act proposal from the state land board and seized on the USGS survey to advocate for the project's development under the newly created United States Reclamation Service. Such a decision meant they would forgo windfall profits and total control of the development and accept in exchange a fustian-like bargain that offered federally subsidized water and engineering expertise as well as technocrats from the East.[7]

The indomitable Camfield and West thus set out again to revive their grand plans. Sugar beet factories that were being built in towns needed the heavily irrigated crop for production. Shortly, federal dignitaries visited the Pawnee Pass area, including Frederick Haynes Newell of the USGS, soon to be the first director of the Reclamation Service. The officials offered optimistic comments about the project. However, in May 1903, engineers under contract with the federal government examined the Pawnee Pass Reservoir proposal and found the notion of a great reservoir west of Sterling unfeasible because of an inadequate and unreliable water supply. Consequently, the GLO threw open to homesteaders the land it had earlier withdrawn at the reservoir site. The Reclamation Service chose to develop its Uncompahgre Project, the prize plum of US Representative John Calhoun Bell of the Western Slope who had advocated passionately for the Carey Act as a worthy experiment eight years earlier.[8]

But Camfield and West had yet another reclamation idea in mind. Their final scheme was to divide the project into separate developments and utilize Colorado's 1901 Irrigation District Act and its 1905 revision. The measure, which Colorado patterned after California's 1887 Wright Act, allowed parties with title to land to form municipal districts to finance waterworks with revenue bonds paid off through assessments on an irrigation district's operation. Many early Wright Act projects in California began failing in the 1890s and left bondholders with little money recoverable on defaulted issues, but later revisions to the law improved the marketability of bonds. The national and international markets for irrigation bonds began with renewed enthusiasm as developers across the West started advancing promotions after 1902. The construction of Camfield and West's lengthy Riverside Canal and the building of the major return flow and storm-water Riverside and Jackson Reservoirs and a string of others off the South Platte thus came to fruition not under the Carey Act but under Colorado's irrigation district law. Camfield became an expert at utilizing this fundamentally important reclamation financing law as the state and the West emerged from the long depression and the sugar beet industry soon dominated most towns in the lower South Platte Valley.

Daniel Camfield died in 1914, his most notable misstep having been his attempt to divert Wyoming-bound Laramie River water in Colorado to one of his few failed projects, the Greeley-Poudre Irrigation District. A visionary, he would have marveled at the massive economic and social transformation supplemental water later transported from west of the Continental Divide would make in developing the lower South Platte Valley. He also would have appreciated how irrigation districts have come to play a vital role in the function of modern water conservancy districts that water interests formed and created—projects such as the United States Bureau of Reclamation's Colorado–Big Thompson Project, which began construction in the 1930s. It is a colossal project undertaken by the Northern Water Conservancy District that cost $162 million and collects hundreds of thousands of acre-feet of the Western Slope's Colorado River water into reservoirs, transports it east through tunnels, and distributes it to reservoirs on the Eastern Slope. The project provides irrigation water to 640,000 additional acres in northeastern Colorado and supplements water supply to nearly 1 million people in towns and cities.[9] All of the development, however, begs unanswerable questions. Would Camfield have imagined the dozens of state wildlife areas scattered along the river's course, some within the reservoirs he built? Or might he have cared about the cost to the South Platte River's natural corridor that has occurred—its controlled flow and its impact on millions of birds and other animals—at the expense of a vision he and West imagined?

One can celebrate or lament or be indifferent to the extraordinary development of the lower South Platte Valley, but modern residents might appreciate the fact that because Camfield and West's Carey Act application stalled for years with the state land board, it did not become mired in the 1894 law's constraints. The reality that this obscure agency made no effort to advance the project's consideration with the GLO may have spared it the doomed fate of every Carey Act scheme in Colorado. Once a company contracted with the state land board to segregate acres, after which the board signed its own contract with the GLO for the gifted land, developers seldom abandoned the Carey Act. In Camfield and West's instance, the state land board and the GLO never executed a contract to segregate land from the public domain for the South Platte scheme.

To understand the Carey Act in Colorado is also to grasp the role the State Board of Land Commissioners of Colorado played in administering it. Beginning with Camfield and West's project, every Carey Act proposal—a total of thirty-four—was subject to the agency's bureaucracy. The board's authority dated to 1876 and resides in the state constitution that created it. When it entered the Union, Colorado (an area of 66.48 million acres) received about 4.35 million acres of public domain land from the federal government in the enabling act permitting admission. The land grants included 3.7 million acres (Sections 16 and 36 of each township) for public schools, 500,000 acres for public improvements, 32,000 acres each for a penitentiary and public buildings, 46,080 acres for saline land, and 46,080 for university and normal school (teacher training) purposes. Congress also granted 90,000 acres for the establishment of the state's agricultural college. The board's constitutional requirement was to generate revenue from the land grants by selling and leasing the land, known as state trust land, with the proceeds going to each grant's purpose. Except for the school sections, the board selected its choice of the other granted lands from the public domain. (Decades later, it selected indemnity lands, specific exchanges for school sections absorbed with the creation of federal forest reserves in Colorado.)[10]

Thus, the state land board specialized in monetizing federally gifted land. As Colorado's population and economic growth accelerated throughout the 1880s, the legislature revised much of the law governing state-owned lands to bolster income flows to the various fund purposes and to more effectively promote immigration to vacant lands across the state. The revisions created the office of a full-time register to supervise the agency's appraisers, surveyors, timber wardens, mining royalty supervisors, stenographers, and other clerks who oversaw Carey Act filings. To fund its administration of the Carey Act, which only began after Camfield and West withdrew their application and other developers began to file as the depression lifted, the board used money collected from the sale of segregated land to settlers. In Colorado and other states the cost was fifty cents per acre—twenty-five cents per acre upfront and the remaining twenty-five cents upon receiving the patent. The board also collected nominal document filing fees from developers. Clerks recorded all Carey Act transactions into ledger books that detailed a company's compliance with each pertinent section of the law. The board also made maps of a company's segregation, recorded the surety bond taken out to complete a project, and compiled other documents such as a company's correspondence with the board. It also maintained custody of settlers' water contracts sold by the developers, as well as their payments on those contracts it placed in escrow at various banks until the board granted permission for construction. Fortunately for historians, these documents after 1902 are voluminous. But virtually no state land board records exist for the Pawnee Pass Irrigation Company or for an ill-defined and quickly withdrawn scheme by Theodore C. Henry in southeastern Colorado north of the Arkansas River in 1900.[11]

Significantly better recordkeeping by the state land board and its stepped-up administration of the Carey Act, perhaps not coincidently, occurred just as the irrigation bond market began recovering from the depression and during the decade after California successfully began rehabilitating Wright Act bonds in 1897. Irrigation bonds were the essential loans from lenders that funded nearly every private large-scale waterworks in the world. Indeed, bonds of all sorts financed nearly every large

American enterprise, from railroads and shipping canals and factories to light and power companies as well as water projects. The marketplace convergence of irrigation districts, privately owned corporate reclamation projects, Carey Act projects, and short memories about investment risk opened the floodgates of the irrigation bond market. Chicago bond firms took particular interest in irrigation bonds, which were generally ten-year or fifteen-year serial bonds (maturing at regular intervals) that yielded 6 percent interest, with principal and interest paid twice a year. Like all securities and equities of the time, no rating agency graded these bonds for risk. Structurally, all that differentiated the bonds was the collateral offered to the bondholder as security. Irrigation district bonds, which were municipal revenue bonds, constituted a lien on the land within the district's taxing authority. Private irrigation companies collateralized their irrigation bonds with liens on their land and, later, waterworks. Carey Act bonds, in contrast, had only the promise of future collateral in the form of a hypothecated bond also backed up with a lien on the future waterworks.[12]

Carey Act lands remained the possession of the federal government until a settler received the patent to the land. Although the United States Congress, in 1896, had authorized states to create a lien on Carey Act lands before settlement (and permitted patent to pass to a state as soon as an adequate water supply was made available), no state provided for making the cost of building the waterworks a direct lien on the lands. As such, there was no authority to make the cost of a lien on the land before construction. Consequently, developers and bond dealers, with the support of state land boards, sought to make the lands a basis of credit without a direct lien on them. To accomplish this crafty maneuver, before construction, developers sold water rights on deferred payments under contracts that bound the purchasing settlers to pledge mortgages on their lands as soon as they received title. Developers then hypothecated the settlers' water contracts, which thereafter became debt notes. These notes, which were the deferred payments on their water rights, were the security for the bonds sold for construction. Hypothecating a settler's water contract meant that a Carey Act company placed it with a financial trust company that also sometimes serviced the bonds. Once a Carey Act settler paid off the debt to the development company with the deferred payments, the hypothecation ended. The trust company then released the water contract, and the state land board issued a land title (patent) to the settler.[13]

When a settler bought a water contract from a Carey Act development company, the individual was assuming a serious debt obligation. The purchase of water was appurtenant (tied) to the land in parcels of no more than 160 acres. Down payment was usually 10 percent, and prices of the water that attached to the land varied from project to project—from twenty-five dollars to sixty dollars per acre. The purchaser might take limited comfort from the fact that a state required that contracts and down payments be escrowed with the state land board until the board permitted a development company to sell its bonds and place the contracts with a trust company. Usually, state land boards required that the dollar amount of notes a company executed be larger than the amount of bonds a company could sell, 125 percent to 150 percent of the bonds in some cases. Nevertheless, even though the Carey Act was administered under federal and state laws, there was no government guarantee of indemnity to the settler purchasing a water contract or the lender who bought a bond if a project failed. If settlers defaulted on their notes, those notes became worthless collateral because title to the land remained with the government.[14]

The reclamation of arid lands was thus a risky business. More accurate, it was speculative, as the terms of its private financing reflected. But even without the specter of marketplace risk, such as projects that began under the direction of the United States Reclamation Service after 1902, settlers faced a harsh reality. By 1920, the total of that irrigated land constituted just 10 percent of all the reclaimed land in the West. As one environmental historian particularly critical of the federal effort has argued, farmers' meager profits left them unable to pay the high cost of dams and irrigation projects. Then, after years of heavily subsidized federal assistance to farmers, the projects (especially in California) eventually fell into the hands of corporate elites and federal technocrats. In any event, the argument of this *hydraulic society* underscores the extraordinary degree to which managing risk and who assumed that risk played out in the reclamation of arid lands.[15]

The fundamental conundrum of reclamation—a question federal officials considered differently than did private developers—was whether desert lands could pay for the cost of their irrigation. Until farmers could irrigate land, that land was uninhabitable. Unlike easily irrigated areas, the cost of reclaiming vast benchlands, while initially low for settlers, bore not only the cost of construction but also the cost of interest upon financing it, as well as the running cost of maintaining the system. Every possible irrigable acre would need to produce a profit to service a heavy debt. For developers, there was not only the task of locating reclaimable land with sufficient water for settlers but also of selling the scheme to financiers. Consequently, the expenditure of very large sums of money was necessary before settlers, who were to repay the cost, located to the land. In addition, there was the matter of developers enticing middle-class settlers who had the farming skills and the capital to create farms from such a wilderness.

As the reader will see throughout this book, those who imagined that their Colorado Carey Act proposals would successfully transform large sections of Colorado's arid wilderness understood the conundrum of financial risk and the detailed process of utilizing as well as amending the 1894 law to fit their needs. Of the thirty-four proposals (see appendix B) developers submitted to the state land board, twenty-eight never advanced beyond the promotion stage. Many of these speculations undertook some financial analysis and hydrological studies, but their promoters endlessly exaggerated the viability of all twenty-eight. So boundless was their promotion of these speculative ventures, regardless of adequate water supply, that the promoters persuaded the Colorado General Assembly to successfully petition the United States Congress to grant Carey Act authority over former Ute Indian Reservation land in 1907 and to grant the state an additional 1 million acres in 1911, as developers had claimed more than the original 1-million-acre grant. All of these grand speculations, some of which proposed the development of hundreds of thousands of acres, failed to generate a single investor dollar and often tied up land that otherwise might have been open to general settlement under the homestead and other laws. Four other speculative projects (three on the Western Slope and one in the San Luis Valley) managed to sell water contracts to settlers and persuaded the state land board and the GLO to segregate the schemes from the public domain. However, these developers never finished their irrigation systems. Unlike two Carey Act schemes that came to fruition in southeastern Colorado, which are the focus of this book, little evidence of any construction remains of the four uncompleted projects. Nonetheless, the rift between developers and settlers there offers insight into how each project ultimately failed.

The uncompleted Carey Act projects west of the Continental Divide and in the San Luis Valley were efforts to transform livestock grazing country that early settlers had been using for decades. The land was once the domain of the Utes, whose various bands had lived throughout the mountains and high valleys of the southern Rockies for hundreds of years. Until the United States Army forcibly removed them to reservations in Utah and southwestern Colorado in 1880 following the Meeker Massacre, they, like their predecessors the Ancestral Puebloans and the Fremont Complex of peoples and the Folsom and the Clovis cultures before them, had successfully adapted to the region's limits but also to its abundance. The flood of white settlers and miners onto Ute land, beginning with the gold rush in 1859, was catastrophic for the tribe, which ceded title to millions of acres of land in successive treaties with the United States. The American ethos of dispossessing Native Americans of their land and exploiting nature for profit, regardless of consequence, was deeply held by Addison J. McCune, who made the first major effort to build a Carey Act reclamation project after Camfield and West's unsuccessful proposal on the South Platte.[16]

McCune knew the Western Slope like few others did. He had worked its mines, surveyed townships, platted towns, built irrigation systems, promoted economic development, and served as Colorado's state engineer. McCune also knew that the greater portion of Colorado's perhaps 15 million acre-feet of annual surface water flowed west of the Continental Divide to comprise the vast Colorado River watershed. He surmised that 10 million acres of undeveloped benchlands and plateaus there were irrigable. In 1903 he left his position as state engineer and, with Grand Junction, Colorado, founder Monroe L. Allison and Denver bond broker William C. Johnston, immediately proposed developing 38,000 acres of benchland above the Little Snake River northwest of the town of Craig under terms of the Carey Act. The men formed the Colorado Realty and Securities Company and anticipated that the Moffat Railroad would be traversing the region. Coalmines would follow to boost population, and the lucrative sugar beet industry might gain a foothold if irrigation were possible from excessive flood flows from Slater Creek, a tributary of the

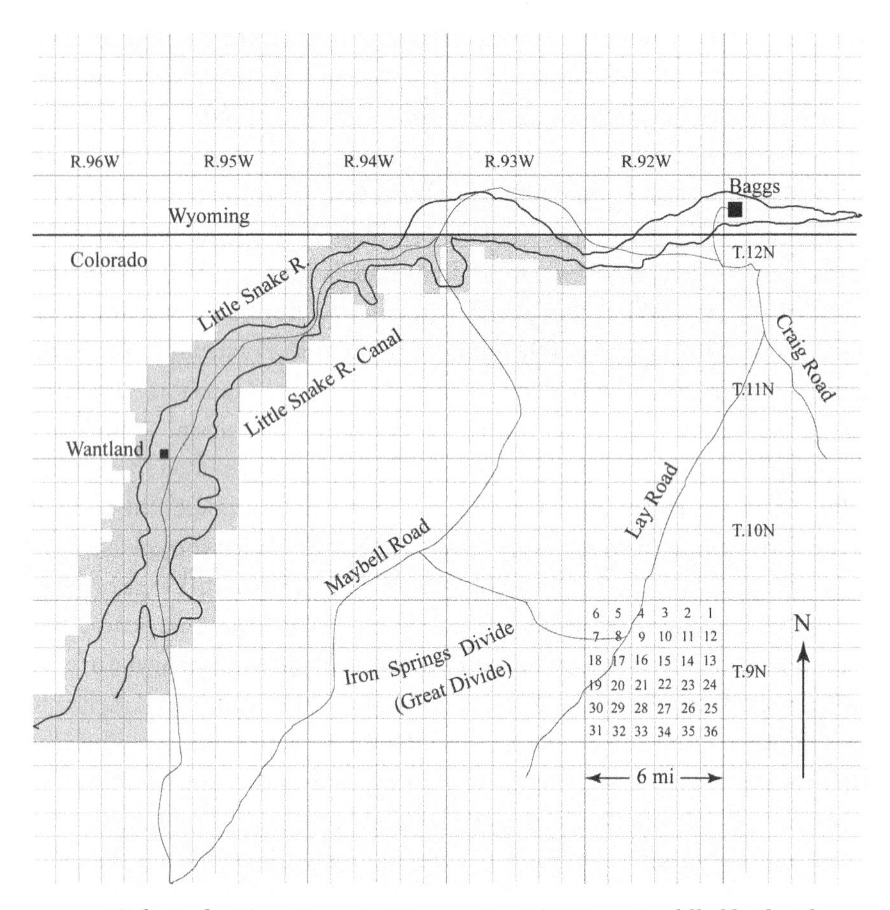

MAP 2. **Little Snake River Carey Act Segregation List No. 3 straddled both sides of the Little Snake River. Its planned reservoir was to be located in Colorado near the Wyoming border, several miles southeast of Baggs.**

Little Snake. McCune planned a 35,000-acre-foot reservoir on the creek and a forty-five-mile canal to the irrigable land southwest of Baggs, Wyoming. The state land board designated the proposal Carey Act Segregation List No. 3; in late April 1903, the board signed its contract with the GLO on behalf of the company, with water rights set at $25 per acre (later adjusted to $45). The company estimated the project's cost at $200,000. But it took until 1905 for the state land board to approve the company's selection of 38,000 acres. The state engineer doubted that the project's water supply was sufficient, but the company resurveyed the reservoir site, showing greater storage capacity. Discouraged investors quit supplying funds until 1907, when McCune and others transferred the company (renamed Routt County Development Company) to the ownership of Denver investors, who finally moved the project beyond the promotion stage.[17]

Promoters extraordinaire headed the new company: Lewis C. Greenlee, retired superintendent of the Denver Public Schools; Warren R. Given, newspaperman turned irrigation bond broker (and later Ku Klux Klan leader); and C. E. Wantland, general land agent with the Union Pacific Railroad and associated with the Denver-based Colorado Land Headquarters, one of the state's largest land settlement companies. Greenlee pitched the scheme to teachers, and Given promoted it to people in the newspaper business—all middle class and white. The company lured hundreds of prospective settlers to inspect the sage-covered lands. By 1908, Addison McCune was back overseeing work on the canal, but no work had started on the Slater Creek Reservoir. By 1911, more than 200 people had put down payments on water contracts, and the state land board permitted the company to issue nearly $500,000 in ten-year bonds at 6 percent interest. The number of hypothecated bonds the company sold is uncertain because company records are nonexistent. Moreover, the irrigation bond market was beginning what would be a historic collapse that made the securities unmarketable. In any event, the company never paid the lenders any principal or interest. Meanwhile, company officials continued work on the canal and laid out the townsite of Wantland. Workers hauled tons of concrete block and lumber to the site. By 1912, the company had exhausted its funds, and all construction on the town and irrigation project halted. Moreover, the company had not even begun work on its proposed Slater Creek Reservoir or another one planned on Savery Creek. The Routt County Development Company, though not technically bankrupt, was a huge liability. After 1914, the state land board returned the segregated acreage to the GLO and returned the down payments on water contracts to the purchasers, none of whom ever settled on the land.[18]

Meanwhile, several miles to the east of the Little Snake Project, also in 1903, the Great Northern Irrigation and Power Company began designing its reclamation plan for the area. The company was an enterprise of David Halliday Moffat, perhaps the richest person in Colorado. The company had unsuccessfully attempted to block McCune and Allison's segregation but nonetheless pressed forward, hoping to develop 240,000 acres

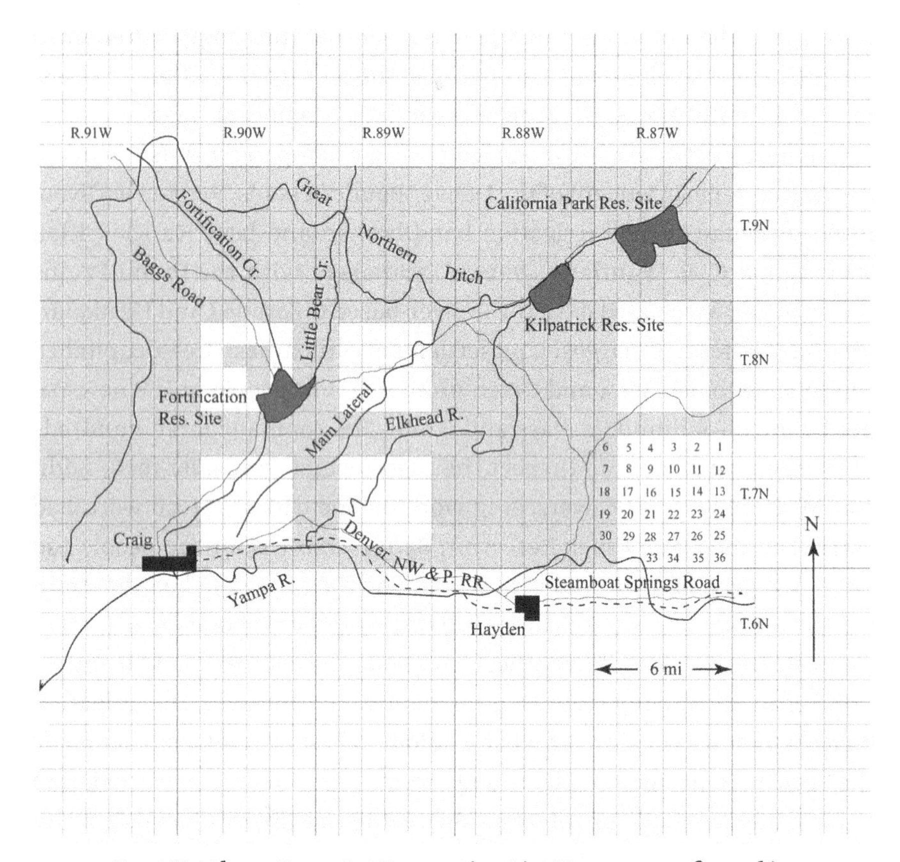

MAP 3. **Great Northern Carey Act Segregation List No. 15, as configured in 1911 from 140,000 acres to about 70,000 acres, was the largest swath cut for a Carey Act project in Colorado.**

of federal and state land board acreage in Routt County. A general optimism about reclamation's potential on the Western Slope may have come from favorable reports by the newly established United States Reclamation Service (USRS), which was conducting its general reconnaissance of the state. The USRS noted possible reservoirs on the Little Snake and tributaries of the Yampa River but chose to build the Uncompahgre Project near Montrose and, years later, the Grand Valley Project near Grand Junction. Surprisingly, Great Northern did not use any of the USRS data to back up its plan, which was to capture runoff from the Elkhead Mountains, move it into reservoirs, and transport it through canals to its proposed land northwest of Craig.[19]

The company's fortunes were dependent on the Moffat Railroad, which was the most expensive railroad construction to date. Following a financial restructuring of the railroad in 1907, Denver lawyer and Moffat confidant Lafayette M. Hughes headed the Great Northern Irrigation and Power Company and in 1909 applied for the massive segregation with the state land board under the Carey Act. The board designated the project Segregation List No. 15 but reduced the acreage to 140,000 acres. A year later the company signed its Carey Act contract with the state, with the cost of water set at $45 and the estimated cost of the project at $1.5 million. The state land board then executed its contract with the GLO. Apparently, Hughes, Denver tycoon Lawrence C. Phipps, and other deep-pocketed investors spent $200,000 on surveys, engineering, and legal costs for the project but no construction occurred. Nonetheless, the company lured more than 100 families to the land with minimum down payments. By 1912, with the irrigation bond market showing no signs of recovery, what had become the largest and most complex Carey Act project had stalled and begun its slow death. The company never issued bonds, and much to the fury of settlers, it successfully petitioned the GLO for extension after extension to complete the system—a provision the 1902 amendment to the Carey Act permitted. Company inaction tied up the vast acreage for nearly a decade while about 100 settlers who had started to develop farms and ranches (as the reader will see in chapter 5) eventually received land patents from the GLO under special dispensation.[20]

Carey Act developers refusing to give up on their schemes despite endless failure and settlers' frustration and despair was a constant theme of the law's use in Colorado. In nearly every proposal, whether the pure promotions that never sold water contracts to settlers or the six projects that did sell contracts, developers refused to voluntarily terminate a scheme. Developers always considered the state land board's designation of acreage—be it a Carey Act project or an option to develop any state-owned land—to be a business asset. The asset calculation, although an intangible one that developers never wanted to give up, took on greater importance to them when the United States Congress amended the Carey Act in 1910—making it possible for a state, at a company's request, to temporarily withdraw land from general settlement and keep interlopers from seeking compensation from developers. In addition, Carey Act developers on the Western Slope, who had successfully lobbied Congress in 1907 to permit development on former Ute Indian Reservation land, were equally assertive that their claim to land was a business asset. However, to those developers' chagrin, Congress reaffirmed a requirement that the tribe was to receive $1.25 per acre sold to settlers, as stipulated in the 1880 Ute Agreement.[21]

Notably, only one Carey Act project on former Ute Indian Reservation land took a proposed development to the point of selling water contracts to prospective settlers. It was the Ignacio Project, whose developers had spearheaded the 1907 amendment to include former tribal reservation land. Lorenzo M. Sutton envisioned irrigating 16,000 acres of the 500,000 acres of the former Southern Ute Reservation the Hunter Act of 1895 had thrown open to public entry (settlement). The law required the band, one of two that remained in Colorado after 1880, to accept its smaller reservation in severality (meaning individual ownership, not communal holdings) and begin farming and ranching. Sutton's plan was to develop an orchard paradise by drawing excess floodwater from the Los Piños River and from floodwater storage in Emerald Lake, forty-five miles upstream. Sutton hired a Salt Lake City land settlement company to promote the scheme and set the water cost at $45 per acre and land at $1.75 per acre ($1.25 to go to the Ute Fund and 50 cents to the state land board). One revealing aspect of the promotion of the scheme was the vernacular and racist rhetoric of F. A. Wadleigh, who had a long history with the Salt Lake City company: "It is an established fact the Indian is lazy. He was too lazy to develop the rich soil of the valley that has long been an unclaimed reward for enterprise." He concluded that the $1.25 per acre to the Ute Fund was justifiable compensation. In all, by 1910, more than 200 people had put money down on 12,000 acres. Meanwhile, the state land board refused to allow the project to issue bonds, and the GLO refused to approve the scheme, citing insufficient water supply. Prospective settlers, understandably irate that no construction had taken place, demanded their escrowed money back from the state land board. At one point, developer Sutton sought a receivership for the project but was unable secure one and persisted with the doomed project. Not until 1917 did the Ignacio Project mercifully end when the state land board withdrew the segregation, returned the settlers' escrowed money, and returned the acreage to the GLO for homestead entry.[22]

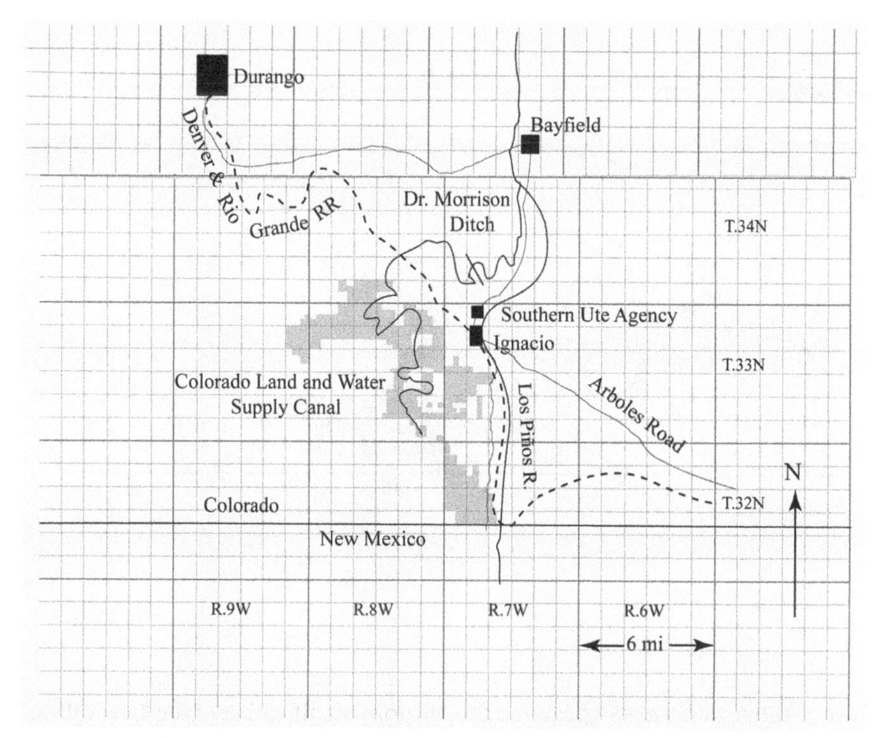

MAP 4. Ignacio Carey Act Segregation List No. 6, situated amid Southern Ute lands, comprised 16,000 acres of benchlands.

The fate of the failed Ignacio Project, like that of the Little Snake and the Great Northern, exemplified the disconnect between private developers' hopes of windfall profits and the reality of unstable financing, economic rollercoasters, and the physiographic challenges of reclaiming high sagebrush lands. The same detachment occurred with the Toltec Project, the fourth Carey Act project that sold water contracts to settlers but never constructed a waterworks. The 1909 scheme of Denver attorney Henry H. Clark, the project sought to revive the old Taos Canal Company plan from the 1890s in the southern San Luis Valley. Clark perhaps thought the project might replicate Theodore C. Henry's great canals on the Rio Grande near Monte Vista. Pioneering Hispanic settlements, whose residents had been successfully irrigating small plots in southern Colorado for half a century, surrounded the Toltec Project. (I discuss this contrasting perception of land use relative to Carey Act development above the Purgatoire River in chapter 4). Clark sought windfall profits on 15,000 acres he

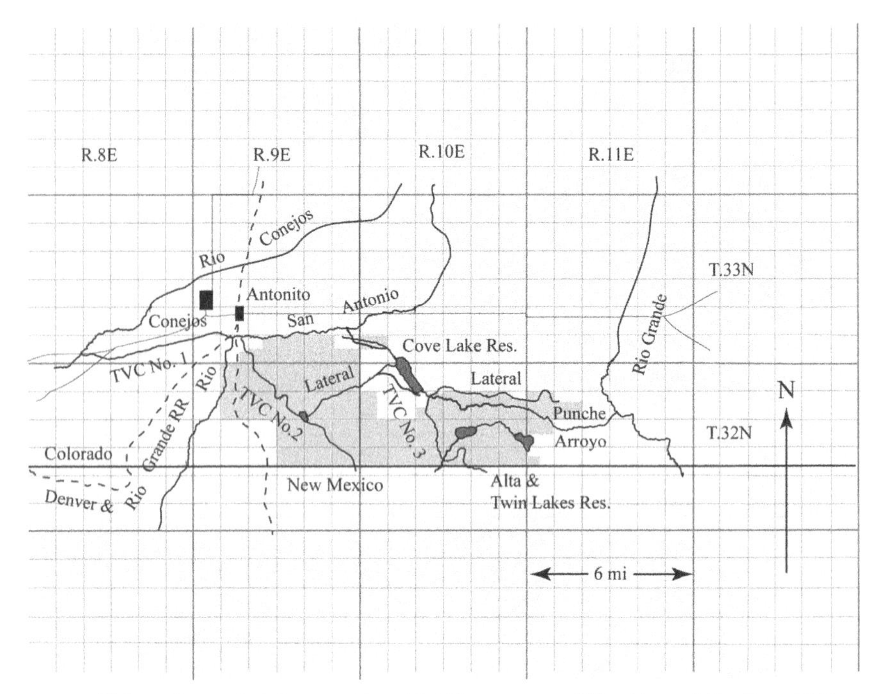

MAP 5. **Toltec Carey Act Segregation List No. 10, a modification of the Taos Valley Canal system, vastly overestimated its eventual 12,000-acre segregation, shown here in 1909. Prospective settlers purchased parcels closer to Antonito and southeast of Cove Lake Reservoir.**

hoped to irrigate with flood flows from the Conejos River stored in Cove Lake Reservoir. More than 200 prospective settlers put money down on forty-dollar-per-acre water contracts on half of the segregation as Clark unsuccessfully sought to sell $3 million in bonds during the collapsed irrigation bond market. In 1914 the state land board canceled the stalled project, and six people who had paid upfront and settled on the segregation received title to their parcels as Desert Land Act patents, as had those involved in the Great Northern scheme.[23]

The Toltec Project in the San Luis Valley and the three Carey Act projects on the Western Slope were spectacular in their failure, but private and federal irrigation development in both regions advanced nonetheless, albeit most often unprofitably for the developer, settler, and investor. As modern historians of the regions have shown, the complexity of irrigation and water development in general across the Colorado and Rio Grande watersheds proved more the product of local involvement than was perhaps the case in other areas of the West. That local resolve took on historic proportions after World War II with the rise of Western Slope conservative Democrat US Representative Wayne Aspinall of Palisade, the powerful chair of the United States House Interior and Insular Affairs Committee (1959–1973). In that role, he essentially directed the federal development of massive reservoirs across the state and the West while thwarting efforts of environmentalists who were advancing new arguments about the ecological damage caused by dam building and who advocated for federal preservation and expansion of national parks and monuments in the West.[24]

Of course, the four uncompleted Carey Act projects were largely inconsequential to the greater dynamic of reclamation in their respective river basins, and their environmental imprint barely scarred the land. Indeed, only the Little Snake and the Great Northern show remnants of the 1894 law. There, during the 1970s—as with wildlife restoration and conservation efforts in the lower South Platte River Basin and across Colorado—state officials began to aggressively manage segments of the failed projects for wildlife conservation, largely on acreage the state land board had retained as trust land and could not sell to ranchers or other parties. Those wildlife conservation efforts expanded on decades of game management occurring on millions of acres of national forests as well as national park lands—public spaces located within Colorado's vertical landscape.

Meanwhile, across Colorado's horizontal landscape, where virtually no public lands like national forests existed, the vast stretches of public domain on the eastern plains in 1900 proved far too enticing for developers to disregard. Available land, though remote, mattered little to more than a dozen developers with grand designs after 1906. Developers with Carey Act plans in far southeastern Colorado desperately believed the region's storm waters might permit the reclamation of benchlands above the lower drainages of the Arkansas, even though those streams were ephemeral. The intermittent-flowing creeks there crossed the heart of Colorado's cattle country—a region the 1894 Frederick Haynes Newell report on John Wesley Powell's congressionally funded great irrigation survey (see chapter 1) described as offering little possibility for irrigation. The report acknowledged the considerable importance of the Purgatoire

River and its irrigation supply at the foot of the mountains but pointedly remarked that "the portion of its drainage traversing the plains did not receive a perennial supply of water and . . . through a great part of the year is completely dry." It noted that except for heavy rains, very little water ever flowed in any of the channels in far southeastern Colorado.[25]

Newell's report about far southeastern Colorado's aridity was apparently unread by the region's residents. In 1906 an enterprising dentist from Rocky Ford began surveying the reservoir location that would become the Two Buttes Carey Act Project. That same year, a young engineer and a young lawyer began mapping the site of the Las Animas–Bent Project, also known as the Muddy Creek Project. Their proposals, ironically in perhaps the state's most arid region, would be the only projects in Colorado that came to fruition under the Carey Act and would lay bare the extraordinary lengths to which developers went in trying to salvage hopeless failures.

3

THE COLONY AT TWO BUTTES MOUNTAIN

Beneath the silhouette of a twin-cone peak that rises 400 feet above the dry plains of extreme southeastern Colorado twenty miles from the Kansas border, the irrigation project that would become Carey Act Segregation List No. 7 began with little notice in February 1906. Dr. William D. Purse, a Rocky Ford, Colorado, dentist and part-time realtor, proposed trapping floodwaters of Two Butte Creek, an ephemeral southern tributary of the Arkansas River. His idea was to make a significant profit by developing and selling an irrigation scheme near the promontory. On the morning of February 16, Purse, age forty-six, bundled in wool and with a small survey crew, began mapping two reservoir sites in a long, deep canyon of the creek that straddles the Prowers-Baca County line about thirty miles south of Lamar and fifteen miles northeast of Springfield.

As Purse and his crew surveyed the nearly dry streambed and mapped the reservoir's sites, they unknowingly stepped across numerous lithic scatters—stone tools and various artifacts left by Indigenous people who had lived in the region for more than 100 centuries. Unknown to Purse,

https://doi.org/10.5876/9781646426492.c003

just west of where he surveyed, the Millennial Site establishes that Native American occupation dates to 7,000 years ago. Conversely, in a matter of just years, Purse's plan would forever transform a landscape on which the Clovis, Folsom, and other peoples who inhabited the continent had lived. The last Native Americans to live in this region perceived the land in a vastly different way and fashioned a way of life unimaginable to the doctor. For them, the land and water were never commodities. The Southern Cheyenne called Two Butte Creek "Mahks' ĭ tsĭ kā' ŏ ĭhka" (Piles of Driftwood). The tribe knew well the stream's flooding tendencies. For generations, the Cheyenne had followed the bison not only for subsistence but also for participation in the robe trade with white traders. Master equestrians, they had also mastered living in the region until the intrusion of whites in the aftermath of the 1859 gold rush, the nearby Sand Creek Massacre (1864), and their removal to reservations in Oklahoma, Montana, and Wyoming. But the Comanche—perhaps the greatest equestrians of all Plains tribes and who played a major role in dispersing the horse, the source of wealth on the Great Plains—dominated the land south of the twin-cone peak and far into Texas. Their economic, military, political, and diplomatic dominance enabled them to thrive in the near desert until epidemics and the United States Army defeated them (and the Kiowa) during the Red River War of 1874–1875 and eventually forced them onto reservations in Oklahoma.[1]

Purse's perception of the land for its commercial value likely blinded him to the meaning of the flint-knapped artifacts he stepped over and even that of the several pictographs Native Americans had marked on the cliff of Two Buttes Canyon. In addition, he would have given little thought to the countless numbers of bison, pronghorns, and other wildlife that once thrived in the region. The creation of private property, not conservation or preservation of animals and landscapes, was his purpose. He located his two impoundments on vacant federal land, although near his second reservoir site was acreage that belonged to the Bent County Bank in Las Animas, which had repossessed failed farmsteads in the region during the 1890s. Only a handful of homesteaders remained in the vicinity. Complying with Colorado law, he filed his maps of the reservoirs at the courthouses in Prowers and Baca Counties and with the state engineer. He also filed for rights-of-way across the public domain with the General Land Office

(GLO). Purse's development plans were formless and not specifically meant to be a Carey Act project. But the filings prevented others from claiming the rights-of-way and gave him time to fashion a scheme utilizing the federal land. That time might also have allowed him the opportunity to turn a quick profit by selling the idea to interested investors and bond dealers, whose marketing of irrigation bonds had begun to take off just as large-scale private reclamation projects were under way in the West.[2]

We know very little about William D. Purse. During his youth he had done surveying in his native Missouri. It is uncertain if he had any business connection with his Rocky Ford neighbor George W. Swink, the pioneering irrigator and farmer whose promotion of irrigated agriculture extended up and down the Arkansas Valley and spurred the development of the sugar beet industry by 1900. Purse was a novice at reclamation, though he had surveyed and claimed rights to the Purse and Medor Irrigation Ditch north of the Arkansas and near Rocky Ford in 1902 that irrigated less than 200 acres.[3]

For a small-time real estate peddler, his survey of the Two Buttes reservoirs was a considerable undertaking. He located his reservoirs twenty miles downstream from where, under the direction of General John Wesley Powell, the United States Geological Survey (USGS) had marked out a reservoir (Site No. 43) during its comprehensive mapping of potential irrigation developments in the arid West in 1889–1890. The great survey, as mentioned in chapter 1, was the United States Congress's momentary funding of reclamation during the cession controversy. The government survey had not only identified irrigation reservoir sites but also marked those sites for withdrawal from the public domain. The government surveyors calculated that the Two Butte Creek watershed was more than 800 square miles of drainage with no significant impoundments. These were the data on which Purse based his premise that reclamation was possible in a region that was otherwise grazing land or dryland farming country.[4]

When Purse traveled south from the well-developed Arkansas Valley town Lamar, located on the Atchison, Topeka and Santa Fe Railroad (AT&SF), he left Prowers County and entered Baca County where no railroad existed. The land had few roads, and none of those that did exist were graded. There were no culverts, bridges, electricity, or telephone lines.

One telegraph line ran into the county, and the few people who lived in the dusty villages of Springfield and Vilas were dependent on mail service for outside communication. The old bypasses of the Santa Fe Trail—the Aubry Cutoff, the Granada to Fort Union Road, and the Cimarron Cutoff running through the southeastern corner of the county—had passed into history. Evidence of the great cattle trails north from Texas—the National Cattle Trail along the Kansas border and the Penrose Trail traversing the county's western edge—had virtually disappeared. The colossal cattle ranches that once dominated the region after the removal of Native Americans—the JJ Ranch on the Purgatoire, the Miles operations on the Cimarron, and the Thomas H. Godwin Ranch on Butte Creek—had all become smaller operations, but they still controlled the seeps and other waterholes. In 1900 only 759 residents lived in Baca County.[5]

William D. Purse had entered an essentially empty county that comprised roughly 2,565 square miles—about 44 miles north to south and about 55 miles east to west. However, a recent exodus of perhaps a few thousand people may have caused Purse to contemplate, if only momentarily, the nature of his development plan. In the mid-1880s a wave of people, all white Americans, began settling in southwestern Kansas. The westernmost extent of that settlement pushed into southeastern Colorado, where the USGS had previously begun surveying the public domain for homesteaders. Several years of very good precipitation seemed to obscure the region's arid condition and to validate the claims of celebrated Great Plains promoters whose propaganda extolled the rainfall, the soil, and other conditions as ideal for intensive diversified agriculture. In southeastern Colorado, speculators organized into town companies and unabashedly inflated the region's potentialities. Between 1886 and 1890, more than a dozen boomtowns sprang up in the heart of cattle country, and hundreds of nesters followed the town speculators onto homesteads across Baca County, which was then in eastern Las Animas County.

Naive hopes and speculative greed combined to rapidly fill the land, but both impulses would also conspire to empty it rapidly. Thousands of farmers arrived by covered wagons. They usually constructed simple basement-like dugout homes and practiced agriculture as they had in more arid regions. Subsistence farmers sold their wheat, corn, and silage to the local market. General goods and building supplies came by wagon from Lamar.

In areas of extreme southeastern Colorado, a homesteader occupied virtually every quarter section of land. The state legislature awarded (and residents later ratified) the distinction of Springfield as the county seat in April 1889 when it created Baca County. At that time, it also created other counties in eastern Colorado that were experiencing an influx of land speculators and settlers seeking autonomy from older counties. Yet the exodus of people from Baca County was remarkably sudden and drastic. Despite constant overtures by railroads and surveys, it was not until 1926 when the AT&SF arrived from Manter, Kansas, that tracts extended there. Lesser precipitation amounts across the entire Great Plains returned to the more usual norm. In Baca County the year 1888 was the driest in twenty years. By 1890 its population was 1,479. Most people had simply abandoned their claims to the government. On one day in 1890, a single mortgage company sold sixty-six farms. Many settlers headed their wagons to the Oklahoma strip, others moved to towns in the Arkansas Valley, and still others returned to Kansas. The few hardy souls who stayed eked out a living by ranching and farming the dry land. In 1894 the extreme drought continued. People who stayed in the county hoped that a second wave of settlers might one day lay claim to the 1.4 million acres of public domain open to homesteaders.[6]

Purse's survey of Two Butte Creek generated excitement among local county residents, but so did the heavens once again. Although three small ditches from the creek already existed near the planned reservoirs, their users irrigated less than 200 acres of pastureland. In September 1904 a torrential flood destroyed the ditches when it hit southeastern Colorado. By 1905, the drought across the Great Plains finally broke. The following year, more than nineteen inches of precipitation fell north of Two Buttes Mountain, and for the next several years steady, timely moisture was the norm. The sites that William D. Purse had surveyed in the winter of 1906 had, like the region, turned to a flowered carpet of yellow prickly pear cactus and thick, deep-green grasses in late April and early May. By midsummer, the cactus blossoms had faded, but even in August the land was still green, belying its usual parched condition. For Purse the moment could not have been more promising. But he also understood the spectacular failure of previous farmers who had followed what they thought was an extending Rain Belt, the myth that farming itself generated precipitation.

Nevertheless, their settlement and that of other residents of Baca County had taken only 13 percent of the available public domain land. In 1907, as the yearly precipitation continued, a second wave of white settlers began moving to farmsteads in eastern Baca County. Some, like those in the first wave, arrived by covered wagon, but others in the second wave eventually came by automobile. Representative of homesteaders were the Osteens, a poor family from Muskogee, Oklahoma, that settled in the far southeastern corner of the county. Dryland farmers, their home was a two-room half dugout, a subterranean excavation they dug that had windows at ground level and a roof. Not all families were as poor as the Osteens, but most settlers soon began utilizing the dry farming methods of Hardy Campbell. His popular journal, *Western Soil Culture*, detailed methods of planting drought-resistant wheat and utilizing deep tilling techniques and extolled the virtues of agrarian life. Across Baca County, a few thousand dryland farmers came, but across the eastern plains of Colorado they came by the tens of thousands.[7]

The second wave of dryland farmers into Baca County dramatically contrasted with Purse's idea of the land's profitability and his belief that irrigation offered a more economically sustainable method of farming. He was not alone in that belief. At a 1907 Trans-Missouri Dry Farming Congress in Denver, Joseph M. Carey, no longer a US senator, spoke to hundreds of attendees and cautiously advocated experimentation in dry farming but frankly acknowledged that he was a wet farmer and more inclined to irrigation's transformative wonder. He warned of the serious consequences that might befall the thousands of rapidly settling farmers if the homesteaders were uneducated about the dry farming process. Only by demonstrating the successful production of crops on dry land year after year, he said, might it be safe to settle people on those lands without fear of the homesteaders starving. Such an experience, he asserted, would lead not only to their overall dissatisfaction but to their dissatisfaction with nature, with government, with currency, and with everything in the country as it had existed during the 1890s when people were drought-stricken and driven out of the arid reaches.[8]

Carey's old confidant Elwood Mead did not attend the congress. But Mead's colleague Frederick W. Roeding of the United States Department of Agriculture's Office of Irrigation and Drainage Investigations

Division did. Mead, who had left Wyoming in 1899 to head the agency, had long since come to believe that hostile federal bureaucrats had rendered the Carey Act ineffectual. Roeding presented the ideas of a maturing Mead, quoting long passages of the engineer's observations about the relation of irrigation to dry farming—commentary the irrigation expert had written for the department's 1905 Yearbook. Roeding offered Mead's practical advice to people intent on dry farming. He emphasized Mead's ethical imperative that such settlement on arid lands was appropriate provided that the settlers could be self-supporting. Quoting Mead he asked, "How might the lands be made to support the largest number of people and give them the greatest measure of human comfort?" He continued, "It is one thing to recognize the advantage of irrigation; another to provide for it." Briefly stated, Mead's notion was that on each 160-acre dry farm, between one and ten acres could be irrigated—enough to give a settler the greatest measure of home comfort. Some experiment station research backed his idea, but in a larger sense his philosophy of irrigation's consequence—cogently expressed in his 1903 publication *Irrigation Institutions*—underpinned his belief that the farm model of the arid West could not successfully exist without some irrigation. Roeding concluded by reading Mead's proposal of utilizing three irrigation methods to augment tillage management and crop selection: "Pumping from soil water or underground streams; storage in small surface reservoirs of storm waters or the irregular flow of streams; and irrigation with flood water whenever it can be had usually in the winter and spring, generally spoken of as winter irrigation."[9]

William D. Purse apparently did not attend the Trans-Missouri Dry Farming Congress or any successive congresses. With profit on his mind, Purse located his reservoirs a mile apart. The upstream Wm. D. Purse Reservoir No. 1, he estimated, cost $100,000 and held 38,000 acre-feet of water. Wm. D. Purse Reservoir No. 2, he calculated, cost $30,000 and held about 8,000 acre-feet of water that was to come from the larger reservoir.[10]

Purse's rudimentary maps sufficed as first filings with the state agencies, but he had no conceptualized plan to develop the reservoirs into a large reclamation project. In late 1907 (perhaps at the urging of Amos Newton Parrish of the First National Bank of Lamar and Lincoln Wirt Markham, Lamar's leading boosters), Purse joined forces with two

Illinois-based bond brokers with offices in Denver—Harry H. Eberle and Lincoln Bancroft—as well as E. B. Miller, whose identity is uncertain. The men formed the Arkansas Valley Land Headquarters Company and set up an office in Lamar, where other desert land irrigation schemes in various stages of development operated. Their purpose was to sufficiently advance the project to the point where they might sell it to a well-capitalized investor.[11]

By June 1908, Bancroft and Eberle had pitched the company's reservoir ideas to the Chicago bond house F. B. Sherman and Company. Fred L. Harris, an officer of the brokerage firm, soon began advising the would-be developers. Harris, in addition to selling bonds, had conducted feasibility studies of proposed reclamation projects in western states for Sherman and other bond houses. When approached about the possible development of the southeastern Colorado reservoirs, he could little have imagined the life-altering course his earthly journey was about to take. Harris was born on July 10, 1867, in the vicinity of Mount Vernon in southwestern Missouri. When he was seven, his parents relocated the family to McMinnville, Oregon, near Salem. He had a limited country school education but was well-read. At an early age, the boy went to sea, visiting numerous ports. Returning from his voyages, he attended Valparaiso Normal School in Indiana, graduated, and worked various jobs in Chicago. He studied law at night at Lake Forest and was admitted to the bar. Sometime around 1896 he went to work for the Chicago bond house of Trowbridge and Company, which handled large issues of irrigation and other municipal bonds. He also assisted with the company's financing of municipal electric plants and consulted on projects on the Nile and Ganges Rivers. By 1903, Harris was traveling to various locations in the West representing the interests of the company, then known as Trowbridge and Niver. His longest stay was in Idaho, where the company assigned him to manage its interests in the massive Twin Falls Carey Act Project. In September 1906, after attending the Fourteenth National Irrigation Congress in Boise, he left Trowbridge and Niver for F. B. Sherman. The firm oversaw numerous irrigation bond issues, including an unsold portion of the Big Horn Basin development of Salon L. Wiley's benchland in Wyoming.[12]

Fred Harris brought quick order and form to the reclamation proposal on Two Butte Creek. Within days he restructured the company as a Carey

Act project. Harris had William D. Purse resurvey Reservoir No. 1, which showed an increased capacity at 42,000 acre-feet. Purse then dropped his idea of creating Reservoir No. 2. At the end of August 1908, Harris had Purse submit the first maps to the state engineer and the GLO that showed a canal system as well as 22,000 acres of federal land for segregation from the public domain and about 2,000 acres of state land. The proposed land for irrigation was about ten miles southeast of Two Buttes Mountain. Simultaneously, Purse and Lincoln Bancroft notified the state land board of their intensions and, with several other men, incorporated the company on August 17 as the Two Buttes Irrigation and Reservoir Company, with a capitalized stock of $600,000 (one share at $25 for each acre of the proposed development). Each man's outlay is uncertain, but it was probably less than several thousand dollars. Theoretically, eventual profit to the company would be the difference between its construction and financing costs and the sale price per acre to settlers.[13]

Before the state land board could request that the United States Department of the Interior segregate the land, state Carey Act regulations required that the state engineer certify the land as arid and judge that sufficient water and storage capacity existed for the project. In early September 1908, William D. Purse submitted testimonials from area residents to State Engineer Thomas W. Jaycox that attested to the creek's frequent floods. He also supplied data that showed that the region's recent yearly precipitation averaged 17.09 inches. Purse acknowledged to Jaycox that the precipitation figures were probably unrealistic and that he had reduced the figure to 15 inches on his map, which measured drainage calculations at 814 square miles. Jaycox was not satisfied with the data on the proposed Two Buttes Reservoir, especially the 42,000 acre-feet figure. In early October, Lincoln Bancroft pleaded with Jaycox to speed up the application process because interlopers and their "agitated intentions" wanted entry onto the proposed land under the Homestead laws.[14]

During the third week of October 1908, tremendous flooding occurred in southeastern Colorado. In the Two Butte Creek watershed, so much rain fell that it would have filled the reservoir in a single day had it been built. The event must have impressed Fred Harris. By late that month he was in Chicago busily seeking capital to underwrite the proposed project. Meanwhile, his colleagues desperately tried to satisfy the state engineer so he would advance their segregation proposal to the state land board. The company's attorney, Granby Hillyer, wrote Jaycox that Lamar's local economy desperately needed jobs and that building up the south country would be of great benefit to the region and the state. Time was becoming critical for the company. On December 15, Jaycox notified the state land board that in his opinion "it is reasonable to expect [that] the flood flows of this creek will continue, that the amount of water that can be impounded will probably fill the reservoir and that an adequate amount of water will be annually assured, which together with rainfall during the irrigation, or growing season, will be sufficient to mature crops." If the company's asserted capacity of the reservoir were correct, he judged that settlers could irrigate 22,000 acres with 1.5 acre-feet of water. Silting of the reservoir, he wrote, "should not be a problem." His approval of the project, he concluded, was conditional on the company furnishing more data about storage capacity.[15]

Notwithstanding Purse's lower-capacity reservoir calculations, the state land board granted the Two Buttes Irrigation and Reservoir Company a temporary segregation of the requested 22,000 acres the company had surveyed. The board then notified the local GLO of its public domain withdrawal request. As required by the board, the company purchased a $15,000 penal bond, which was 5 percent of the estimated cost of construction and was nominally meant to ensure that the company would complete its contract with the state. The bond was also meant to instill a general sense of confidence for potential investors in the project. Meanwhile, company officials satisfied State Engineer Jaycox's storage capacity concerns by presenting an amended map of the reservoir showing newfound coves and low points missed during earlier mapping, for a revised capacity of 42,000 acre-feet of water. The project remained with the GLO until April 9, 1909, when the state land board signed a contract with the United States Department of the Interior, thus formally segregating the land.[16]

On June 12, 1909, the Two Buttes Irrigation and Reservoir Company signed its Carey Act contract with the state land board. The company agreed to reclaim 22,000 acres of land with the price of perpetual water rights to settlers at $35 per acre (a rather substantial increase from the planned $25-per-acre cost), with payments to begin when the company completed the waterworks. In addition, the board reserved 1,900 acres of

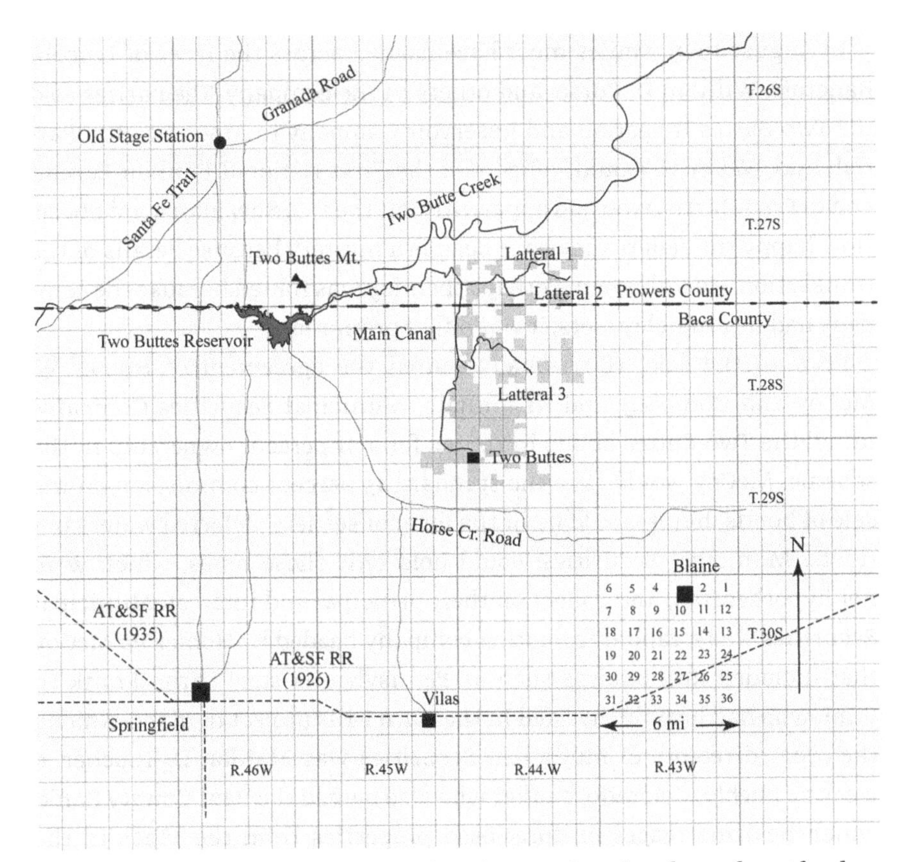

MAP 6. Two Buttes Carey Act Segregation List No. 7 imprinted on a desert landscape that covered nearly two townships, around 40,000 acres. Lightly shaded areas are parcels the GLO patented to the State of Colorado, after which the state transferred some titles to setters.

nearby state land for the company at $5 per acre at a future time. The company was to complete the project within five years and to bring the water within 1.5 miles of a segregated parcel. Until settlers paid their water contracts in full, during which time the company managed the system before turning over the waterworks, annual maintenance costs could not exceed 80 cents per acre. The state land board designated it Carey Act Segregation List No. 7. The company's development plans surely added to growing enthusiasm in southeastern Colorado and especially in Baca County about the potential growth that might complement the increasing numbers of dryland wheat farmers filtering into the region.[17]

In August 1909, smiles must have flashed across the faces of Lincoln Bancroft, William D. Purse, and others in the company. The parties sold the Two Buttes Irrigation and Reservoir Company to Amos Newton Parrish and Welley C. Gould, officers of the First National Bank of Lamar, and to Fred Harris, who had apparently left the F. B. Sherman bond house. How happy the sellers may have been is a mystery because details of the transaction are unknown. Perhaps their joy was muted because the company had no tangible assets; it was only a promotion.[18]

Prior to the sale, Harris had secured the agreement of Homer W. McCoy and Company and its affiliate, Municipal Securities Company, to sell the Two Buttes bonds in return for a 51 percent ownership in the scheme. McCoy would take the standard 15 percent commission on the sale of bonds, but it would be in the form of settlers' deferred water contracts. Municipal Securities would hold only those notes, which were not hypothecated, and collect on their principal and interest. Municipal Securities was a capital formation company headed by Homer W. McCoy that included millionaires such as Pennsylvania steel baron Frank H. Buhl, who had heavily invested in the Twin Falls project in Idaho. Among the other directors of Municipal Securities was Mahlon D. Thacher, a wealthy Pueblo, Colorado, banker who also owned the Bent County Bank, which held mortgages on foreclosed properties from the 1890s in the vicinity of Two Buttes. McCoy and Municipal Securities contributed little upfront money—perhaps a few thousand dollars—as had Parrish, Gould, and Harris to cover purchasing the Two Buttes Irrigation and Reservoir Company and planning the project's development.[19]

Amos Newton Parrish became president of the Two Buttes Irrigation and Reservoir Company, Welley Gould was secretary, and Fred Harris assumed the dual role of company treasurer and general manager. With cash on hand from the deep-pocketed backers, in September 1909, Harris flew into action. He immediately hired one of State Engineer Jaycox's former deputies, Charles Worth Beach, who had trained at the University of Illinois, to design the waterworks and oversee construction. He also hired the Denver engineering firm John E. Field, Abraham Lincoln Fellows, and Michael Creed Hinderlider as consultants. Beach estimated engineering and construction costs at $325,000. By early 1910, a small army of construction workers was throwing dirt.[20]

Fred Harris modeled the development after the Twin Falls Carey Act Project but in miniature and without a railroad, only a hope that one might eventually come. In addition to reclaiming the arid acreage with an impressive waterworks, he meant to build a town where the canals terminated, at the south end of the segregation. To promote the Two Buttes Carey Act Project, he put Parrish to work on a multi-page brochure advertising that 11,000 acres of farmland was for immediate sale and that 11,000 acres would be available for public drawing on October 21, 1909. The brochure highlighted details about the irrigation project and offered data on agriculture in the Arkansas Valley. Included were photographs of plump cantaloupes, huge sugar beet fields, and stacks of alfalfa grown under ditches near Lamar. It claimed that the Two Buttes region received between fifteen and seventeen inches of annual precipitation and quoted engineer Abraham Lincoln Fellows: "1 1/2 acre-feet of water should insure a reasonably good crop practically every year."[21]

Moreover, the booklet proclaimed, "This tract is nearer the Kansas City and Chicago markets than any other Carey Act project which has ever been opened for entry." Water rights on irrigable land, it stated, could be had for a fraction of the cost of established irrigation acreage. Slick and exaggerated, the brochure was typical of the blatant and sometimes shameless booster literature meant to lure farmers to many regions of the West. However, unlike some Carey Act promotions, it did not suggest that the state in any way indemnified the settlers. Harris also broadcast news of the town site's opening in newspapers across Colorado and the Midwest. He hoped to appeal to the specific economic and social class of settlers—white men and *women*—proponents of the Carey Act believed would settle on segregated lands: moderately prosperous, industrious, often temperance people, with sufficient financial means to build farms and the willingness to believe in technology's promise to reclaim desert-like land.[22]

As early as October 1909, the Two Buttes Irrigation and Reservoir Company opened its second office, a makeshift building at the proposed townsite where the company had purchased a quarter-section from the GLO. For those interested, the company ran wagons and automobiles from Lamar to the townsite and to surveyed farm parcels. Remarkably, some people purchased water contracts on parcels from 40 to 160 acres unseen.

For the farmland itself, a settler paid 50 cents per acre (25 cents upon entry, 25 cents upon final proof) that went to the State of Colorado to help fund its cost of administering the act. The state also charged a $1 filing fee and $1 for a certificate of location. When the settlers bought a water right, they were signing a serially maturing water contract with the company for the purchase price of $35 per acre. In addition to the upfront 25 cents, settlers paid a $5 down payment per acre and agreed to ten yearly installments of $3 per acre plus 6 percent interest, with the first deferred payment to begin on October 1, 1911. Early sales totaled more than 7,250 acres. At the well-advertised land drawing on October 21, the company sold more acreage. The next day, the company held a drawing for town lots. Twenty people put down payments on lots totaling slightly more than $2,000. Not every town lot sold. One person inquiring about establishing a liquor store received a polite "no thank you" from Fred L. Harris. The grand two-day production, complete with a huge steam tractor ready to clear fields, helped the company sell more Carey Act parcels.[23]

Those who purchased water contracts did so for a variety of reasons. They were white middle-class men and women willing to commit themselves to an indebtedness of from roughly $1,500 to $6,000. Unlike what the reader will see regarding settlers at the Muddy Creek Carey Act Project in chapter 7, no detailed firsthand account has emerged of what motivated individuals and families to buy into the Two Buttes scheme or that expounds on life there. However, settlers surely took the decision seriously. The names of men dominate the signed water contracts. But often their spouses signed a separate water contract on an adjacent parcel, revealing a collaboration that went beyond the single contract that men alone signed. Even without those contracts, it would be presumptuous to think that women had no role in chancing such a dramatic decision to invest their future in the Two Buttes Project. Moreover, at least nine women who had no spouses bought into the project. One, Anna B. Loomis, stayed with the scheme for years, perhaps hoping for a return on her investment and for the chance to pursue a meaningful life on her own terms.[24]

In the meantime, the state land board judged that only 18,857.14 acres of the 22,000-acre segregation were capable of being irrigated because some sections of land were beyond the reach of the canal system. The company, however, still held its option on 1,900 acres of nearby state-owned land that it valued at $20 per acre. Harris, Parish, and Gould sold the company's last allotment of Carey Act land by early 1911 and retained the option on state land for sale later. The water contracts totaled $660,240, pledged by settlers to the project. Settlers' down payments on water contracts amounted to nearly $100,000, money the state land board held in escrow at the First National Bank of Lamar but began releasing to the company almost immediately when the board permitted construction to commence in late October 1909. At the same time, the state land board began releasing settlers' water contracts to the company for hypothecation, as the contracts it had sold met the necessary state requirement of 125 percent of contracts sold to bonds issued.[25]

All the while, Parrish, Gould, and Harris readied the company's financing bonds for sale by Homer W. McCoy and Company. They calculated that it would cost $475,000 to construct and finance the project. Harris drafted contracts for two issues of ten-year, first-mortgage, 6 percent coupon bonds: one Series A of $365,000 and one Series B of $110,000, dated July 1, 1909. Like irrigation district bonds and general corporate irrigation bonds, note denominations were $500 and $1,000, amounts that permitted the individual investor as well as financial institutions to purchase them. Moreover, the lower entry price made irrigation bonds easier to sell. The bond and water contracts accompanied a first-mortgage deed of trust that Two Buttes Irrigation and Reservoir gave to Chicago Title and Trust and Harrison B. Riley on the company's assets, which at the time were the settlers' water contracts. A fiduciary, Chicago Title and Trust and Harrison B. Riley took the customary 10 percent par value for hypothecating and servicing the $475,000 in bonds. Meanwhile, the cost of Homer W. McCoy and Company's deferred 15 percent commission for the sale of the bonds was a debit to the Two Buttes Company of $65,068. Thus, the project carried an immediate debt burden of 25 percent. Moreover, settlers, the only source of revenue generation, had obliged themselves to a 6 percent yearly bond interest attached to their payments on principal. Indebtedness defined the project, as it did for many new reclamation projects undertaken during the boom years 1906–1914 as developers sought to store the floods of the West.[26]

In February 1910 the state engineer approved plans for the dam's construction. But the company's work on the canals was well under way, with the Leonard and McDowell Construction Company of Lamar cutting the ditches. Work on the canals progressed efficiently under assistant engineer Jay Vandemoer. Crews blasted tunnels in sandstone outcroppings, plowed and scooped earth, and installed three seven-foot-diameter wood siphons thousands of feet long. Men and teams of horses cut more than forty-five miles of ditches. To build the dam, the company hired Denver contractor Dennis Gibbons. But work on the dam stalled, so the Two Buttes Irrigation and Reservoir Company fired Gibbons and undertook the work itself. It was a huge job. Fred Harris and chief engineer Charles Beach managed the effort. Perhaps to the annoyance of teetotaler Harris, Beach judiciously dispersed whiskey to the more than 200 men on the job. At least 150 teams of horses and mules also labored. A large construction camp at the dam site was complete with cooking and dining facilities, a bunkhouse, and a commissary. There were corrals and feed bunks for the livestock. Workers used two graders and forty-six dump wagons during the dam's construction. Common laborers earned $2.00 per ten-hour day; drivers, $2.25; carpenters, $3.50; drivers and horse teams, $4.00; cooks, $3.00; and water boys, 75 cents plus board.[27]

Like nearly every other large dam constructed on the Great Plains between 1900 and 1920, the Two Buttes Dam is a homogeneous earth dam—an embankment built of a single earth material, in this instance, sandy clay scraped from below the surface soil from an area north of the dam. Workers first built a concrete core that was anchored to each end of the rock walls of the narrow canyon on Two Butte Creek. Laborers carried the excavated earth in two-wheel, horse-drawn wagons and spread it with scrapers. With a mule pulling a roller made of railroad car wheels on an axle, they compacted the soil. Workers rolled the material in layers about twelve inches thick an average of six times and lightly moistened each layer with water pumped from the small stream. They placed a partially steel-reinforced concrete slab (later covered with hand-placed riprap when the slab slipped) on the upstream embankment. Workers blasted rock to construct the outlet tunnel at the north end of the dam and installed head gates that controlled the water's flow in the canal. The dam extended 1,500 feet along its crest and measured as high as 108 feet in one spot. Workers located the reservoir's spillway at the dam's south end.[28]

Significant problems plagued the construction. The summer of 1910 was extremely hot. In July, twenty days had temperatures of 100 degrees, and ten days reached 106 degrees. Dust made conditions miserable for the workers and the horses. Laborers left the jobsite in droves, which delayed the work. In August it rained 4.5 inches in one hour, causing serious damage to the ditches and to one siphon on a ditch. Engineers estimated that about 5,000 acre-feet of water escaped. Repairs and finishing work on the canals caused further delay. On September 24 the dam was three-fourths completed; on November 26 the company declared the system completed. The cost of construction, though ongoing, stood at $268,200.[29]

The dam created Two Buttes Reservoir, the irrigation supply for the segregated lands. At capacity the impoundment could seemingly hold 42,000 acre-feet of water. Under the best circumstances the "duty" of water was about 1.5 acre-feet per segregated acre. Nearly all the supply would come from summer floods that might rush into Two Butte Creek along its fifty-mile course from the spot where it originated in western Bent County. The project's main canal from the reservoir fed the system of canals that delivered north-flowing water destined for the Arkansas to the east and then south to the segregated Carey Act lands. In early 1911 chief engineer Charles Beach reported to the state land board that the company had expended $317,000 constructing the waterworks.[30]

By February 1911, about fifty families had arrived at Two Buttes. They came with enough money to build their cabins, sink domestic water wells, and prepare their farming operations. As quickly as they could, the settlers carved ditches from the company's main canals to their fields. By late April, they were waiting for water. But the company could only provide a small distribution of water because inadequate rains and no flooding had occurred. During the entire month of June, the reservoir was empty. Men, unable to farm, hurried to nearby towns to get work to sustain their families. In July the reservoir held less than 1,500 acre-feet of water, and the company opened its first run of water in the main canal. Fred Harris notified the state land board that the company was prepared to supply water and that it expected all those who had purchased acreage to make their yearly installment payment. Infuriated settlers threatened not to pay and immediately protested to the state land board. Edward Keating, a member of the board, traveled to Lamar and met with the protesting settlers, who explained that it was too late for planting and that the call for payment placed an undue burden on them. Keating, a former journalist and progressive, sympathized with the settlers and persuaded the company to delay collecting payments until the following year. Technically, the company had built the system, and the situation of too little water was a circumstance not of its making.[31]

Throughout 1912, precipitation amounts, at fifteen inches, averaged just above the region's norm, but again no flash floods occurred in the Two Butte watershed. Inflow to the reservoir measured 7,319 acre-feet. Water ran in the ditches on forty-three days, and row crops did well but only on less than 2,000 acres. Moreover, much of the crop experienced heavy hail damage. Few of the settlers working irrigated or dry land could make their October payments to the company. The year 1913 proved even more disastrous for the settlers. Less than 2,000 acre-feet of water flowed into the reservoir, and precipitation averaged only 13.58 inches. Crop production was minimal, as the company ran what water it could for just twenty-seven days. Again, most settlers found it impossible to make payments. For a second year, the Two Buttes Reservoir and Irrigation Company defaulted on its bonds.[32]

At the behest of Homer W. McCoy and Company, Fred Harris pled the case to the bondholders and Chicago Title and Trust and Harrison B. Riley that more time was needed to make the enterprise profitable. He persuaded them to accept a postponement on all payments of principal for seven years. As had become the favored method of many Carey Act companies, the Two Buttes Irrigation and Reservoir Company granted time extensions for payments by executing supplemental contracts with settlers. These contracts, numbering in the dozens at Two Buttes, allowed settlers to pay interest only for the first seven years after the initial down payment. The remaining payments on the principal were to begin in the seventh year and to be made by settlers in annual installments for the next ten years.[33]

The year-after-year circumstance of the company's often dry reservoir, the plight of the settlers, and the default on bonds coincided with intense state and national scrutiny of the Carey Act's general failure to reclaim any substantial acreage of arid lands in the West. As will be seen in chapter 5, critical examinations by the State Board of Land Commissioners of Colorado played out with transformative importance to the state's other

Carey Act projects—especially the Muddy Creek undertaking in adjacent Bent and Las Animas Counties. The United States Congress, responding to various complaints about Carey Act failures and in response to the 1911 irrigation bond collapse, began examining the law in the autumn of 1912. The congressional inquiry tapped data compiled by the GLO. Charles W. Wells, the agency's Carey Act investigator in Colorado, reported very favorably on the Two Buttes venture despite its small amount of water and the fact that settlers were irrigating only a fraction of the acreage.[34]

But the Two Buttes Carey Act Project had critics. In September 1913, James U. Harris, an appraiser for the state land board, called the Two Buttes irrigation project "one of the biggest frauds in the history of Colorado." He asserted that if the settlers were to be protected, the state would voluntarily have to place water on the land at its own expense. *Denver Farm and Field*, an agricultural weekly whose editor complained that Carey Act bonds had significantly contributed to the worldwide collapse of irrigation securities, editorialized that the project was a fraud and that the 22,000 acres could never be adequately irrigated. He wrongly contended that the state had given the impression that it stood behind the projects and was thus obligated to fund the irrigation project if the settlers were ever to receive patent to the land.[35]

As if sent from Heaven, significant snow fell in December 1913, and timely rains fell throughout 1914. At Two Buttes Reservoir almost twenty inches of precipitation transformed the parched region. Reportedly, 40,000 acre-feet of water filled the reservoir. In June 1914 the state land board notified the settlers that actual residence on their entries must begin immediately and must continue until they had made final proof to the state for patent under the Carey Act. To meet the burden of proving up (the parlance of acquiring title to the land), the entry person was to irrigate or otherwise cultivate at least 20 acres of every 160 acres within three years. Settlers only irrigated about 3,000 acres (21 percent of the segregation's farms), but the timely rains alone watered much of all the crops grown that year.[36] Meanwhile, the state land board readied its petition for the United States Department of the Interior to patent the segregation to the State of Colorado for the land's eventual transfer to settlers. How much acreage it would seek to have segregated was uncertain, but the amount was nowhere near the original 22,000 acres contemplated by developers.

An undercurrent of pessimism also began to form among the project's otherwise limitless-thinking founders. In 1914 Fred Harris wrote to Dr. William D. Purse and drew an effective conclusion about the colony's beginning. The doctor—who, with his wife, Etta, had purchased separate parcels—had made no effort, like many others, to prove up their land by cultivating at least one-eighth of their claim after two years. Dr. Purse had fallen on hard times and was now a salesman for a boiler company in Kansas City. Harris wrote: "The proposition has been a heavy looser to us all. It is extremely doubtful if we will ever be able to pay off our bonded indebtedness, to say nothing of nearly $100,000 of floating debt. I wish I was merely broke, for then I could recover, but I am saddled with a personal debt which will take years to liquidate. No doubt I could buy more than one half of the segregation for $250 for a quarter, and assume payments of interest, principal and maintenance." Two years later, in June 1916, Harris again wrote the doctor, who wanted to sell his parcel: "We have a long drawn out and hard proposition here, and I have been staying with it faithfully in the hopes of working it out. But it is a killer and I am about to the limit of my staying powers. We are so far from a railroad, and there is so little hope of one in the near future that good farmers keep away, and we have to get along with the inferior class, as a rule. We have had no rain here since last August, and you will know how dry it is." But optimistically he concluded, "Yet our crops are doing well, especially alfalfa, and prices are satisfactory. We have about 45 feet of available water in the reservoir, but there has been no increment this year. . . . This drouth [*sic*] may (and is) forcing drylanders to look up to us."[37]

In the coming years (as the reader will see in chapter 8), Fred Harris—and the colonists—would have much more to concern themselves with than impressing the neighbors as they sought to salvage the project. Meanwhile, fifty miles to the west, the Muddy Creek Carey Act Project was under way and about to give additional meaning to the boundless minds of developers.

4

THE BIRTH OF THE MUDDY CREEK CAREY ACT PROJECT

Fifty miles west of Two Buttes Mountain, on a wide plain of the Purgatoire's benchlands south of Las Animas, Colorado, Will R. Murphy, a twenty-five-year-old civil engineer, set out in 1906 to resurvey two irrigation proposals that had been dormant for years. Murphy, a native of Bent County, knew well the country's varied landscape, and he took as a certainty that capturing the region's flash floods could sustain a colony of farmers across the Heinan Flats ten miles south of his hometown. He and his pioneering family had been part of the Arkansas Valley's agricultural development from the time the United States Army conquered and displaced the Cheyenne and Arapaho to the supplanting of Hispanic subsistence settlements along the lower Purgatoire River to the dominance of livestock grazing and the rise of industrialized sugar beet farming. Murphy partnered with Cecil E. Sydner, a thirty-one-year-old attorney and longtime Las Animas resident, to handle the necessary state and federal legal filing requirements for developing the land, most of which was public domain. By utilizing the science of hydraulic engineering and

https://doi.org/10.5876/9781646426492.c004

the dynamic of capitalism, the men had high hopes for substantial profits from reclaiming grazing land by building a waterworks on ephemeral streams adjacent to the Purgatoire. Limitless though their thinking was, they could never have imagined how their plan would morph, how other developers would reinvent their plan, or that it would become the most contorted of all the Carey Act projects in Colorado.

The two young men, without a second thought about the ecological wonder of the vast and unique Purgatoire country, envisioned reclaiming about 38,000 acres nine miles south of Las Animas that was above the Purgatoire River's reach by irrigation ditch. The last of the bison there had died decades before, but some wild horses remained, as did mule deer, mountain lions, bears, and various songbirds. For decades, Hispanic settlers living along the river's abundant course from Trinidad to its confluence with the Arkansas River near Las Animas regularly consumed nearly all of the Purgatoire's annual flow of perhaps 100,000 acre-feet. Murphy and Sydner planned to reclaim their selected land by capturing floodwaters and winter runoff into two separate dams—one on Smith Canyon, a tributary of the Purgatoire, and one on Muddy Creek, a tributary of the Arkansas. Just as William D. Purse had done, the men utilized data gathered by the 1889–1891 United States Geological Survey (USGS) exploration that segregated reservoir locations from public entry. That survey identified a potential reservoir on Smith Canyon Creek (Site No. 40) and one on Muddy Creek (Site No. 41), which the survey identified as a western branch of Rule Creek.[1]

Murphy and Sydner were not the first to see in the great national irrigation survey the possibility for widespread irrigation on lands south of the lower Arkansas. In 1902 the irrepressible Theodore C. (T. C.) Henry submitted plans to the state engineer for his "Great" Chaquaqua Irrigation System—a massive canal similar to his 110-mile Fort Lyon Canal and his Bob Creek Canal north of the Arkansas. The Chaquaqua Canal would head in the south bank of the Arkansas seven miles east of Las Animas and run east, zigzagging to the Kansas state line; it would be supplied with additional water from USGS-identified reservoir sites that he proposed developing on Chaquaqua, Smith Canyon, Muddy, and Rule Creeks. (Interestingly, his scheme made no plans to impound Two Butte Creek.) Other irrigation promoters dreamed; Henry fantasized. Henry envisioned that the project's development would occur under Colorado's newly enacted Irrigation District Law of 1901 and would include public domain with private land. But Henry took no action to advance his great scheme, perhaps because his reclamation plans were incomplete and lacked available financing. Other developers later proposed unsuccessful efforts to segment the vast scheme as the Bent and Prowers Irrigation District. Meanwhile, Murphy and Sydner carved their own project across multiple townships from the same Chaquaqua scheme that included a feeder canal from Smith Canyon Dam to Muddy Creek Dam and thence a supply canal to the segregated lands.[2]

In snapping (transferring) their irrigation scheme to the American survey grid system, Murphy and Sydner, like other optimistic developers in the region and in the San Luis Valley, were effectively supplanting decades of Hispanic customs of colonial settlement patterns and irrigation traditions. Since the 1840s, Hispanics (in this instance people of Mexican ancestry) had begun settling on Mexican land grants in what would become southern Colorado, establishing the San Luis Valley villages that included Conejos and San Luis. In the Valley of the Purgatoire, the first *pobladores* began settling in the early 1860s. Profoundly different worldviews separated early Hispanic residents and developers such as Murphy and Sydner and the likes of Henry H. Clark, whose Toltec scheme (mentioned in chapter 2) was imprinted on Hispanic Colorado. As historian Donald J. Pisani put it: "The basic purpose of Mexican law was not to stimulate private enterprise but to irrigate the maximum acreage. . . . Wherever possible irrigation should be a community endeavor."[3] Whereas townships, ranges, and sections allowed for the commodification of land, private property, and profitability, Hispanic settlement customs emphasized communal values that included private and public ownership of land.

Across Hispanic Colorado's well-watered floodplains, cultural values anchored settlers not only to one another but also to the land. The small groups of families from parent villages in New Mexico who settled in these northern settlements spawned their own satellite communities in the region. The new settlers built traditional adobe or stone homes, plazas, and individual gardens; shared community grazing and wildlife hunting areas; and carried forward the communal nature of living in a land that required irrigation for farming. *Acequias*, community irrigation ditches under the administration of an appointed *mayordomo*, were the fundamental feature of water sharing in the settlements. Hundreds of *acequias* have sustained life in Hispanic Colorado; in the lower Purgatoire Valley, below Baca Plaza (later the market town of Trinidad), the simple ditches served numerous communities: Chaquaqua Creek, Minnie Canyon, Cordova Plaza, Nine Mile, Boggsville, and some whose names are lost to time. Beginning in the 1860s, some of these families took parcels under the Homestead Act and variously grew corn, hay, squash, and beans and grazed sheep and cattle. What foodstuffs the residents did not consume they marketed to the growing railroad and coal town of Trinidad or to bustling agricultural towns in the Arkansas Valley where rails brought white residents, who by 1900 had become the majority population group.[4]

Will R. Murphy's life coincided with that of Hispanic pioneers in southern Colorado, but it diverged in profound ways. His father, John A. Murphy, the child of Irish immigrants, had been a soldier under Kit Carson's command at Fort Garland in the San Luis Valley during the 1860s and settled in Las Animas when railroads began laying tracks along the Arkansas River in the early 1870s. The elder Murphy purchased the *Las Animas Leader*, through which from the 1880s to the 1920s he and family members extolled the virtues of economic development. Mainstream racial tropes of the time were prevalent in its pages, but more often it simply ignored the Hispanic experience in the region. For example, the nearby Hispanic community of Boggsville, one of southeastern Colorado's first non-military settlements that dated to 1862, received little attention in the *Leader* for its cultural vitality but rather for its development of the livestock industry by residents Thomas O. Boggs and John Wesley Prowers, both of whom had settled on lands that belonged to their Indigenous spouses.[5]

Boggs, an early employee of Bent's Fort, and his wife, Rumalda, a member of the Cornelio Vigil family, were the first to settle at Boggsville. The land was part of Rumalda Boggs's 2,040-acre share of the land grant parceled out to Vigil's heirs following his death. Prowers, another Bent employee, and his Cheyenne wife, Amache Ochinee, constructed a home near the Boggs family. Between 1866 and 1873, dozens of other families located to the community. The famous fur trader William Bent, nearing the end of his life, established his cattle ranch near the Purgatoire's confluence with the Arkansas. His spread was a 2,085-acre tract of the

Vigil–St. Vrain Land Grant awarded to him and other heirs following the death of his elder brother Charles, killed with Cornelio Vigil in the 1847 Taos Pueblo Uprising. Frontiersman Kit Carson and his wife, Josefa Jaramillo (Rumalda Boggs's aunt who also had claim to the land grant), moved to the community with their children in 1867. Months after their arrival, Josefa died in childbirth, and Carson died soon after at nearby New Fort Lyon, a United States Army compound built in 1867 near the Bent Ranch. Boggs and Carson ran livestock mostly on their spouses' claims south of the Arkansas, and Prowers ran his immense herds of cattle across the vast area north of the river, primarily on claims the United States had awarded to Amache and several other survivors of those murdered in the Sand Creek Massacre in 1864. By 1870, Boggsville was a burgeoning settlement linked to Fort Lyon and early Las Animas. Not only was Boggsville a social center but its farmers supplied the fort and surroundings with foodstuffs. For a brief time Boggsville served as the county seat when the territorial legislature created Bent County from Pueblo County and later expanded Bent County by including parts of Greenwood County, which was the former Cheyenne and Arapaho Reservation abolished in 1874. By 1880, Boggsville was a multicultural community of Anglos, Hispanics, and Native Americans who had blended into the greater economic and social development of the Arkansas Valley from Pueblo to the Kansas state line. Boggsville's population then dwindled, as Las Animas and other towns steadily grew along the railroad where farmers established hundreds of irrigated farms. Across the sweep of arid lands south of Boggsville, sheep ranches expanded, and huge cattle operations came to control millions of acres of public domain across the Purgatoire region.[6]

Boggsville contracted as Will R. Murphy came of age. His grand reclamation idea, beyond its potential for personal profits, was to carry forward the social and economic transformation of the Arkansas Valley. Murphy, after graduating from public school in Las Animas, studied engineering at the University of Colorado and the University of Kansas. For a short time, he worked on engineering projects in Texas that included railroad construction before returning to Bent County to open his own firm and occasionally help with the family newspaper business.[7] His reclamation scheme to irrigate 38,000 acres, some of which abutted Boggsville, would imprint on about 4.5 million acres of undulating and open grazing country south of the Arkansas and east of the Purgatoire, half of which was public domain available for settlement to homesteaders. A contraction of the open-range cattle industry to somewhat smaller operations, a dispirited throng of settlers who fled dry parcels during the 1890s, and the lack of an adequate transportation network in far southeastern Colorado contributed to the undeveloped condition of the arid land. Like the Two Buttes developers, Murphy and Sydner saw in the Carey Act an opportunity to extend irrigation's reach to land seemingly useful only for grazing and dryland farming. Neither man, however, concerned himself with the fact that more than 100 water decrees issued by district courts on the Purgatoire had allowed its depletion, the crux of hydrologist Frederick Haynes Newell's 1889 report that the river mouth often ran dry.

On horseback and by wagon, Murphy began surveying the southwestern component of the project in December 1906. He mapped Smith Canyon Creek, an ephemeral forty-five-mile tributary of the Purgatoire River in Las Animas and Otero Counties that empties floodwaters near the headquarters of the JJ Ranch, an English and Scottish syndicate that operated the Prairie Land and Cattle Company. From the mouth of the drainage south to its deep, rocky, juniper-covered canyon's beginning, Murphy measured the length of Smith Canyon Creek's watershed. He paid special attention to USGS Reservoir Site No. 40, about seventeen miles upstream from the Purgatoire. The government survey estimated that a reservoir there might hold 34,230 acre-feet of water and that mostly public domain surrounded the 1,400-acre site. Whereas the government irrigation survey had merely proposed a storage site for the Purgatoire-bound water, the Las Animas group proposed utilizing a gap in two mesas seven miles northeast of the site to transport the Smith Canyon–impounded water to the Muddy Creek watershed five miles east. Murphy's maps were rough, but he noted the canal's route through Muddy Gap and plotted check dams at various creeks and arroyos that ran into Smith Canyon Creek.[8]

In the summer of 1907, Murphy surveyed the northeastern component of the project. He mapped a large area that included parts of southwestern Bent County along Muddy Creek and its southern tributary, Johnny Creek, which headed near the start of Two Butte Creek. He decided that rather than locate the Muddy Creek Reservoir exactly where the USGS had selected Site No. 41 (a 1,500-acre site with a capacity of 32,780 acre-feet), he would instead locate the dam two miles upstream where it cut between two low mesas. He calculated a reservoir capacity similar to what the government had determined. As with much of Smith Canyon country, nearly all the land around the withdrawn Muddy Creek Reservoir site was public domain. The only privately held land consisted of widely scattered ranching homesteads, bank-owned land, and several parcels owned by the Prairie Land and Cattle Company. The 38,000 acres that he and Sydner envisioned as irrigable on the Heinan Flats and near Boggsville belonged to the federal government. They named their venture the Las Animas–Bent Irrigation Project. Sydner began the process of filing requirements with the state engineer, the state land board, and the General Land Office (GLO) for rights-of-way to the future water storage and canals across the public domain.[9]

Murphy and Sydner's idea to irrigate the lands in southern Bent County, like William D. Purse and Fred L. Harris's plans to water Prowers and Baca County lands, rested on an optimistic but misguided belief that southeastern Colorado's yearly thunderstorms, as well as the severe floods that seemed to occur frequently, might supply irrigation reservoirs. Indeed, in 1904, just two years before Murphy mapped the Smith Canyon component, massive southeastern Colorado floods, which also roared down Two Butte Creek, caused serious damage along the Purgatoire—inundating Boggsville, the similarly diffuse community of Nine Mile, Las Animas, and down the Arkansas River into Kansas. Perhaps this flood factored into Murphy's judgment about designing the scheme. In any event, the 1889–1890 government survey had measured the Smith Canyon Creek watershed at 250 square miles and the Muddy Creek watershed at 140 square miles. Murphy calculated the area of the watersheds to be 450 square miles and judged annual precipitation across that area to be almost fourteen inches by using the triangulation method of the time, which relied on historical data from the United States Weather Service. He used precipitation averages over a ten-year period from the nearby towns of Hoehne (east of Trinidad), Rocky Ford, Las Animas, and Lamar and the Baca County towns of Vilas and Blaine. Those data suggested a runoff of 56,000 acre-feet, about 1.5 acre-feet of water for each acre of 38,000 segregated acres they believed were irrigable. Meanwhile, irrigators tapping water from the Arkansas River were putting 2 to 3 acre-feet per acre on crops.[10]

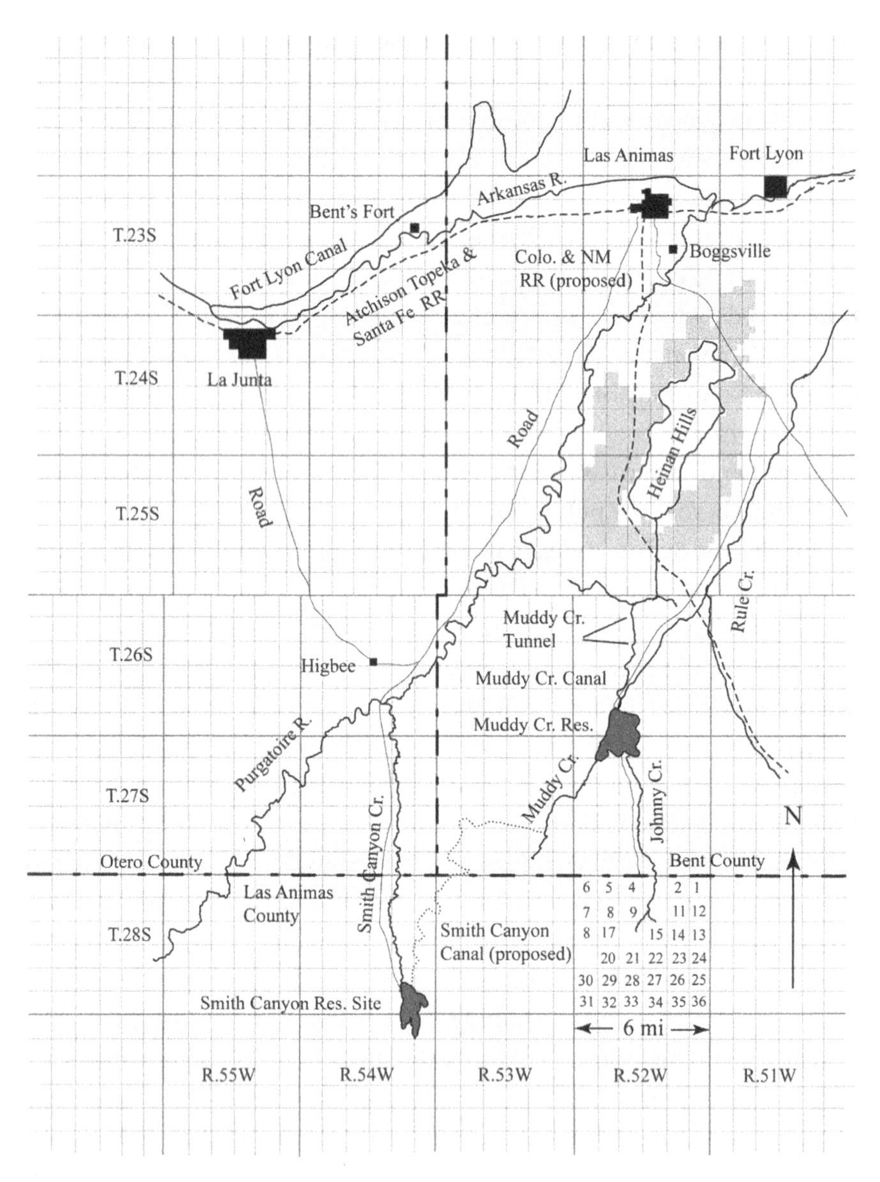

MAP 7. Muddy Creek Segregation List No. 11, with its complex design to reclaim land surrounding the Heinan Hills. The lightly shaded sections represent its selection of irrigable acreage.

Murphy's calculation of the area's historical rainfall and its potential to accumulate storm water was a seriously flawed method regardless of its standardized use by engineers. The error would have serious consequences, similar to the miscalculation by the engineers who built the Two Buttes irrigation project. One environmental historian calls the general misperception the "Chimerical Vision," the product of an unchecked imagination common among Progressive-era engineers. The profession had yet to develop engineering's modern hydrographic applications—formulas that calculate runoff coefficients, peak discharge basin runoffs, and other data essential to scientific assessment. The early hydrologic engineers who designed such remote reservoirs paid little attention to the observations of other scientific disciplines. Geologists, for example, had begun to reveal more telling insight about the nature of present and past environments of the Great Plains, the region between the 100th meridian and the Rocky Mountains. Since as early as 1892, eminent geologist of the Great Plains Louis E. Hicks had cautioned engineers and others who proposed irrigating from storm-water reservoirs. A student of the world's most renowned geologist, Louis Agassi at Harvard, Hicks had made his career by studying ancient environments from the plains of North Dakota to Oklahoma. To the audience, he spoke about the limitations of such irrigation reservoirs: excessive evaporation, extreme variations in rainfall, loss to seepage, and the ratio of catchment basin to reservoir surface (his hypothesis purported that even if rainfall exceeded fifteen inches every year, the success of water storage sufficient for irrigation was slight). He concluded: "Water storage upon the high mesas of the treeless belt is, if not wholly a delusion, at least somewhat delusive."[11]

Engineers at that time were not opposed to the opinions of geologists and agricultural scientists; they were simply inclined to exclude judgments from outside the engineering field. They had undeniably transformed the industrializing nation in profound ways. Western water engineers saw the genius of their work as fundamentally advancing and modernizing US society. Will R. Murphy, as closely as any other young engineer during the Progressive era, witnessed firsthand the social, economic, and political transformation irrigation engineering might bring to a community.[12] He had seen commercial irrigation and mutual companies build the hundreds of miles of canals and dozens of small reservoirs that transformed the lower Arkansas Valley. In the irrigated Arkansas Valley, tens of thousands of acres of alfalfa, hay, melons, and the introduction of sugar beets comprised most of the crops grown. The construction of beet factories in the valley beginning in 1900 infused large amounts of capital into the region and boosted farm and non-farm income. Beet company officials contracted with farmers annually to grow large acreages of the irrigation-dependent crop. By 1902, more than 1,000 irrigation systems included 3,000 miles of main canals and several reservoirs that served 6,480 farms growing sugar beets and other crops. The valley's irrigation systems, if ever able to operate at full capacity, had the potential to irrigate more than 388,000 acres. East and west of Las Animas, Arkansas Valley settlements sprang up: Avondale, Fowler, Manzanola, Rocky Ford, Sugar City, Swink, La Junta, McClave, Wiley, Lamar, Granada, and Holly. Each of the communities had grown into active towns dependent on the economies of bustling railroads, cattle ranches, sheep operations, and irrigated agriculture. Moreover, Murphy had seen in neighboring Las Animas and Huerfano Counties the rise of industrialization with the dominance of coal mining. In Pueblo, the steel industry interlaced with the coal industry, both enterprises part of the Colorado Fuel and Iron Company empire.[13]

Murphy perceived his world through the eyes of an engineer. When he helped out with the family newspaper, that voice often added to boosting the town's economic development. Murphy lacked the national standing of the irrigation movement's most celebrated crusaders, William Smythe and George Maxwell, but he was equally adamant in his belief that transforming a pioneer society and its economy in an arid region could only be accomplished with intensive irrigation. His idea was that the extension of irrigation's reach beyond the Arkansas Valley proper to the benchlands would bring in people of greater financial means and superior character to the common homesteader chancing the heavens with dryland crops. Murphy's newspaper and engineering work gave him an elevated standing in the community, and in 1907 he accepted the position of Las Animas city engineer. He quickly assumed a prominent role in the region's economic development organizations: the Las Animas Commercial Club, the Arkansas Valley Commercial Association, the local chapter of the Commercial and Good Roads Association, and Republican Party politics.[14]

By 1908, Murphy's acquaintance with business and social interests extended across the state. Among those connections was a group of Denver businesspeople eager to join the speculation frenzy that had arisen around irrigation projects and the potential profitability from marketing irrigation bonds. Whether Murphy and Sydner meant for the investors to underwrite the project and play a passive role or intended to sell the venture outright to members of the business community is uncertain. However, during the first year of their association with the Denver group, both men actively continued working on the venture and in June received notice that the GLO had accepted their first filing of documents to develop reservoirs. Heading the group of investors were Dr. Alpheus L. Pollard, a fifty-one-year-old homeopath turned coalmine speculator, and Herbert L. Peebles, a forty-three-year-old shoe salesman turned real estate agent. In the late 1890s, Pollard moved from Anamosa, Iowa, to Denver and began to speculate in developing coal-rich lands on the public domain in northwestern Colorado. The historical record reveals nothing about Peebles or other early investors.[15]

During the summer of 1908, the Denver investors eagerly inquired about the project's substance with Colorado state engineer Thomas W. Jaycox, who offered them a favorable, albeit cursory, opinion based on Murphy and Sydner's sketches of the reservoir system. Pollard and Peebles hired engineer Frank H. Whiting to represent their interests and to closely examine the project. Whiting was a prolific irrigation developer and the designer of the unsuccessful Bent and Prowers Irrigation District project, which had been the original idea of T. C. Henry when he proposed the Chaquaqua Canal system. Its grandiose purpose was to supplement water for farmers who contracted with the American Beet Sugar Company to supply its Lamar factory and one under construction in Las Animas.[16]

Whiting had trained in engineering at Iowa State University in Ames and came to Denver in the early 1890s. He went to Alaska during the gold rush and returned to the Centennial State by 1900. He was the designer of irrigation systems throughout the West. Whiting traveled to Las Animas, inspected the proposal, and reported favorably on the project's feasibility. He estimated the cost of the Las Animas–Bent project to be $850,000. In late 1908 the Denver investors formed the Valley Investment Company as a Colorado corporation at a capitalized stock value of $500,000. Alpheus Pollard was the company's president and Herbert Peebles its secretary. The price for which Murphy and Sydner sold the venture is unknown. But like the Two Buttes Project that Dr. William D. Purse and others sold to

Fred L. Harris and his associates, the project was only a promotion with no tangible assets beyond claims to water rights and land development filings with state and federal agencies.[17]

Insight into Sydner and Murphy's abrupt exit from the venture might be gleaned from Murphy's personal life. During the autumn of 1908 Murphy's wife, Maud, developed rapid-onset tuberculosis, known then as quick consumption. They had met in Lawrence, Kansas, and were married in 1905. Two children were born to the couple: Marilla Frances in 1906 and Maxson Brown in 1907. The national tuberculosis epidemic had led physicians and other practitioners to establish sanatoriums in Colorado and other western states in the belief that dry air was salubrious and might heal the afflicted. Where Maud Murphy might have taken treatment is unknown, but it is possible that she suffered at home. In Las Animas in 1906, the abandoned Fort Lyon had been transferred to the United States Department of the Navy and converted to a makeshift sanatorium. As city engineer, Murphy directly assisted the development of the medical complex that grew to encompass 1,000 acres. And in October 1908 he must have found it especially difficult to separate the professional from the personal. On the twenty-sixth of that month, Maud succumbed to the tortuous disease. Suddenly, Murphy was alone with sole responsibility for two-year-old Marilla and one-year-old Maxson. It was the beginning of a long discontent that would affect his professional career for a decade. He took a more active role with the *Leader*, serving as its full-time editor until 1918, when he again returned to engineering. His friend and cohort Cecil Sydner continued to practice law in Las Animas.[18]

On the Sabbath eight days prior to Maud Murphy's passing, a drizzling rain began to fall in southeastern Colorado. During the night, the wind shifted from the east to the south, an unusual direction for storms in the region. At Holly, near the Kansas border, flashes and streaks of lightning illuminating the western sky indicated a significant storm somewhere nearby. The next morning, the Arkansas from the mouth of the Purgatoire to the Kansas boarder was in a flood state. During the period October 19–21, the river was in a continuous flood state from Las Animas to the Mississippi River. Unbeknown to Arkansas Valley residents at the time, the flood had two distinct storm centers. The first was north of the Arkansas in the vicinity of the Great Plains Water Company reservoirs that supplied lands north of Lamar under the extensive Amity Canal. The second was south of the river, along the Purgatoire and other southern tributaries of the Arkansas.[19]

The southern storm dumped heavy rains south and southwest of Las Animas. Trinidad only received about half an inch. But on the lower Purgatoire and its tributaries, the rainfall was substantial. In Smith Canyon the runoff was of epic proportion. And on Muddy and Rule Creeks, the first important Arkansas tributaries east of the Purgatoire on its south side, the waters rose to an unprecedented stage. J. D. Rhoads, a thirty-five-year resident whose ranch was ten miles above the mouth of Rule Creek, stated that the 1908 flood was several feet higher than he had ever experienced. Along both streams, the flood washed out embankments. Farther east, along Caddoa Creek and Two Butte Creek, the flooding continued. Down the Arkansas the flood destroyed railroad tracks and bridges, haystacks and granaries, head gates and canals, and roads and bridges. Great piles of debris were heaped along the flood's path. In all, the torrent caused more than $250,000 in damage and took the lives of several (specific number unknown) people.[20]

Despite its ferocity, the October 1908 flooding differed little from previous episodes. The intensity of rainfall was probably greater than usual in the area where Murphy and Sydner had proposed their waterworks, and overgrazed conditions there may have permitted more rapid runoff. The Atchison, Topeka and Santa Fe (AT&SF) incurred significant damages east of Las Animas. Its tracks crossed the flooding tributaries of the lower Arkansas. Damage from cloudbursts along these streams was not uncommon. For Alpheus Pollard and other investors of the Valley Investment Company, it must have been a moment of heightened optimism. USGS officials investigating the aftermath of the flood noted that the event illustrated the possibility of additional storage facilities that could provide for the irrigation of a significantly larger area of the region. The investigators further concluded: "Floods of this character can be prevented only by constructing storage reservoirs on the tributaries. Until this is done life and property will never be safe in that section."[21]

Pollard pushed forward with developing the Las Animas–Bent Irrigation Project. In 1909 he managed to persuade H. D. King, the owner of a small coalmine near Aguilar north of Trinidad, to become an officer of Valley

Investment Company. Joining them in the irrigation venture was J. W. Mullins, whose identity is a mystery, and George E. O'Brien, a Denver grocer. King, Mullins, and O'Brien brought just enough cash, about $40,000, to keep the irrigation plan afloat in the hope of its eventual financing through the irrigation bond market. The parties presumed that if all of Valley Investment's 38,700 acres sold at $60 per acre and generated $2.32 million, a profit of $1 million to them might be possible after expenses.[22]

During early 1909, engineer Frank H. Whiting refined Valley Investment's water storage scheme. In addition to developing the specifications of the company's two major reservoirs and the main canal systems, he proposed two smaller reservoirs with feeder ditches along Smith Canyon Creek and at Long and Spring Canyons. On the main canal from Muddy Creek Reservoir to the segregated lands, he planned a 1,400-foot tunnel and four small reservoirs with feeder ditches to the main canal. Thus, with the object of capturing every freshet in every canyon, arroyo, draw, and rivulet, he sought to capture the maximum volume of all precipitation for irrigation of the segregated lands.[23]

In May 1909 the officers of Valley Investment filed their plans with the state engineer and applied to the state land board to request that the GLO make a preliminary segregation of lands under the Carey Act. On November 29, 1909, state engineer Charles W. Comstock reported his investigation of the project: "I am convinced that there is sufficient unappropriated water available to the Valley Investment Company for the irrigation of 24,000 acres of land, and I know that the proposed reservoirs, canals, and other works are adequate for the storage, distribution, and utilization of this water. I therefore recommend that this segregation of 24,000 acres be made as requested." Two weeks later the state land board made its application for the permanent segregation of the lands to the GLO. In approving the project, however, the two state agencies restricted the segregation to only 24,000 acres—14,000 acres less than proposed—but agreed to the cost of water delivery to the land at $60 per acre. The illusion of a $1 million profit suddenly became $600,000 if development occurred under the best of circumstances.[24]

As the Las Animas–Bent Carey Act Project slowly made its way through the various layers of state and federal oversight, the company's officers learned how much of the project's design was beyond their control and how national reclamation policy, still in its infancy, might foretell the future of irrigated agriculture in the arid reaches of the West. In early 1909, Pollard and his associates discovered how fiercely competitive water ownership rights in Colorado were. On Rule Creek, the drainage into which Muddy Creek flows, Las Animas residents Frank W. Foote and Sydney Flinn had claimed water rights in 1907, just months after Murphy and Sydner's state water filing on Muddy Creek. The Bent County Reservoir Company proposed building a 50,000-acre-foot-capacity reservoir and ditches to develop about 20,000 acres of farmable land adjacent to Rule Creek, southeast of Las Animas. The company filed a protest with the Colorado state engineer asserting that Valley Investment Company's claim to Muddy Creek's water threatened to render inadequate its supply of water to develop farms. Although the state engineer had no power to forbid someone to claim the right to water, he did have the authority to judge the feasibility of Carey Act projects. However, in late 1909, State Engineer Charles W. Comstock rejected the protest and ruled that sufficient water existed to irrigate both projects.[25]

The state engineer's decision, beyond its local significance that each company possessed the right to use water from each watercourse, underscored the administrative and legal custom in Colorado to allow water to be over-appropriated. In this instance, in the state's already most over-appropriated region—the southeastern corner, Irrigation Division No. 2—Colorado's enshrined water law, rooted in the Doctrine of Prior Appropriation and adjudicated in courts of authority when disputes occurred, endowed first users with a permanent right to water as long as it was put to beneficial use. Moreover, the law, having also established usage rights in order of priority date, established that those users with a senior priority right held substantial advantage over those with a junior right—a significant consequence during times of drought for downstream users such as the Bent County Reservoir Company. Of course, the very senior users in the Arkansas Valley proper and along the Purgatoire had rights that predated the Muddy Creek and Rule Creek users by as much as forty years. To protect their water assets, the lower Arkansas users had formed the Arkansas Valley Ditch Association, which played a principal role in the complicated matter of the Arkansas River's use in the United States Supreme Court's *Kansas v. Colorado* decision in 1907. In addition to rejecting federal control over interstate streams, the Court established the "doctrine of equity" whereby each state held an economic stake in the river. Not until 1949 was there a resolution between Kansas and Colorado, when both states agreed to the Arkansas River Compact.[26]

In the meantime, Valley Investment Company's reclamation plans sat with the GLO while its field engineer for southeastern Colorado, Charles W. Wells, analyzed the project's feasibility throughout early 1910. On June 18, 1910, Wells delivered a favorable report about the project. He concluded that Smith Canyon and Muddy Creek would supply ample water to each of the company's two planned reservoirs to irrigate 24,000 acres. Moreover, he recommended that 404.07 additional acres be approved, given company engineer Whiting's design of small dams along Smith Canyon Creek and those along the lower canal system. He provided an itemized list of the project's probable cost at $892,800, a figure he considered conservative but that matched Whiting's estimate. Wells believed the annual maintenance (the cost to carry water to the intended settler) should not exceed sixty cents per acre. The secretary of the interior thereafter approved the plans and signed articles of agreement with the State of Colorado on October 1. Finally, on December 10, 1910, the officers of Valley Investment Company signed a formal contract with the state land board—fifteen months after the state had submitted its segregation application to the GLO. H. D. King and J. W. Mullins soon left the company, and Alpheus Pollard resumed his role as president with Herbert Peebles and George E. O'Brien staying on as directors. Frank H. Whiting became the company secretary. The esteemed engineer's role in the development increased even as he designed other irrigation projects in Colorado and Wyoming.[27]

The new board of directors of Valley Investment first advertised its scheme at the Eighteenth National Irrigation Congress held in Pueblo in September 1910. Hundreds of delegates attended, including the country's leading reclamation voicers—Frederick Haynes Newell, Joseph M. Carey, forest conservationist Gifford Pinchot, and representatives of the steel, railroad, and sugar beet industries. Perhaps Pollard, Peoples, and O'Brien thought they might find investors among the throng, but they found none. But this congress was one of the more consequential of its annual gatherings. Its delegates shifted the organization's support of state control of interstate streams, which favored industries that profited from federal lands, to an endorsement of federal control of waterways by the United States Reclamation Service, which favored the supporters of federal reclamation development and forest conservationists. While the Pueblo gathering signaled to the United States Congress a consensus within the reclamation movement's most prominent organization, Valley Investment Company's directors likely saw the sheer largesse of the federal program—especially its funding mechanism—as contrary to the free-market values they embraced.[28]

The Newlands Act of 1902, which created the Reclamation Service within the United States Department of the Interior, launched the federal government's direct attempt to settle farmers on the land and construct waterworks. In addition to the service's wide authority to decide a federal reclamation project's location, character, and feasibility, it administered the act's unique funding mechanism. The service paid for the initial construction of federal projects from the sale of public lands, the proceeds of which went into a revolving reclamation fund. To replenish the fund, the service imposed on the settler interest-free user fees to be repaid over a ten-year period for the cost of water, thus supplying the revolving fund for future projects. But by 1910, repayment to the fund by settlers, many of whom were in default, seriously lagged behind federal development. In June 1910 the United States Congress provided a loan to the service for existing projects by authorizing the issue of $20 million in United States Treasury–backed bonds. The Department of the Treasury at that time sold bonds at par (face value), and the debt instruments were virtually risk free. This issue was of five-year bonds that yielded 3 percent interest per annum, half the yield of the usual privately issued irrigation bonds without discounting for broker fees.[29]

The federal reclamation program had been in financial trouble for years as settlement lagged behind the completion of projects. Its continued funding was the subject of serious national policy debate. There was even sentiment to privatize the program. It would take thirty years of future remedial legislation by Congress to spin out the nature and course of federal reclamation. In contrast, the plight of private reclamation in 1910 was dependent on the irrigation bond. Simple in its various forms but risky in its return, that bond had also begun to falter.

By November 1910 and into early 1911, the financial speculation bubble that had inflated the irrigation bond market with exaggerated claims of profitability began to collapse. Of the scores of new private irrigation projects, irrigation districts, and Carey Act projects that had opened across the arid West, numerous schemes remained unfinished. And many of those that had been completed, such as the one at Two Buttes, could not generate the necessary revenue stream to service bond debt. Thus, the worst fears of bond investors quickly materialized as farmers defaulted. Investors holding each type of irrigation bond—private, corporate, irrigation district, and Carey Act—began to fear serious losses. Lenders holding the abstraction of security in "hypothecated" Carey Act bonds must have gasped at their potential losses, given that the bonds held only the illusion of collateral. The spectacular irrigation bond market plunge brought down two of the largest brokerage houses that peddled Carey Act bonds—the Chicago firms of Trowbridge and Niver and the Farwell Trust. In both US and European bond houses, irrigation securities found very few buyers no matter the project's merit.[30]

In Colorado the collapsed irrigation bond market paralyzed new private development of the state's more arid reaches for more than a decade. Colorado state engineer Adelbert A. Weiland recalled years later that this period of explosive speculation in land and irrigation exceeded by nearly every measure the reckless speculating that surrounded the financial bond markets during the state's mining booms: "Promoters, aided and fostered by certain classes—boomers, real estate men and so-called colonizers—loaded many acres of Colorado's lands to the hub with worthless irrigation securities."[31]

Undeterred, Alpheus Pollard and the other officers of the Valley Investment Company pressed on. Promoters promote. They cast worry aside, believing that other avenues for financing the proposed works were available. Apparently, they sought out German investors, but none lent them any money. Had any bond house agreed to purchase the bonds, the amount of bonds sold would have to cover the $850,000 construction cost, and the market for a bond would have been at a price to the company of 80 percent to 85 percent of par value. Such a price would make the cost of borrowed money to the company 7 percent or more based on the customary 6 percent bond. That cost would have required the redemption payment for the bond to be 20 percent to 25 percent greater than the original proceeds the bonds generated. Nonetheless, the company continued to advertise the project and marked out the dams and canals and individual parcels of land for sale to settlers. By late 1912, the company had spent more than $40,000 surveying lines and building 3.5 miles of the Smith Canyon Canal. Eventual profitability, the company's directors continued to maintain, would come from selling water contracts to settlers at $60 per acre.[32]

But the company could interest no one in the project. It could not sell a single share of water or acre of land, and it did not secure a dollar in financing. Rather than fold the company and lose what personal money they had in the promotion, Alpheus Pollard, Herbert Peebles, and George O'Brien decided to wait for a recovery in the crisis condition of irrigation financing. The wait would prove to be in vain.

5

"LAND BOARD, MAKE THEM FISH OR CUT BAIT"

Free Carey Act land was a developer's delight, stalled and failed project after failed project notwithstanding. In early 1911 the State of Colorado doubled down on its commitment to developers and the Carey Act. On behalf of unabashed developers who refused to accept limits, the general assembly sent a memorial to the United States Congress and the president requesting a gift of 1 million acres more of public domain. Earlier, the states of Idaho, Wyoming, and Nevada had successfully secured additional Carey Act lands by amendment, and Colorado lawmakers followed that course. Colorado's request included a letter from the State Board of Land Commissioners of Colorado stipulating that across the state, applicants had filed on all of the original 1 million acres plus 232,000 more. The General Land Office (GLO) had approved 259,000 segregated acres that comprised the Little Snake, Ignacio, Two Buttes, Valley Investment, Toltec, and Great Northern Projects. The remaining 980,000 temporarily segregated acres were awaiting approval by the GLO or action by the state land board. The letter emphasized the urgency of the request and

https://doi.org/10.5876/9781646426492.c005

noted that developers were seeking even more acreage every day. It did not, however, mention that Colorado had just restructured the state land board and given it broader powers, a reform that would accelerate the acceptance of new Carey Act lands but eventually exert greater oversight of all projects.[1]

Representative Edward T. Taylor, a second-term congressman from the Western Slope, and Senator Simon Guggenheim of Colorado introduced the measure into Congress, arguing that the request would expand business opportunities. Each man sat on his respective chamber's Public Lands Committee and secured a supporting letter from Acting Secretary of the Interior Samuel Adams. The bill quickly moved through Congress, with unanimous consent. President William Howard Taft signed the amendment to the Carey Act law permitting Colorado the withdrawal of a second million acres of federal arid land on August 23, 1911. Almost seventeen years to the day had passed since the Carey Act became law in 1894, and Colorado had yet to patent a single acre of the desert land to a settler.[2]

Conventional western boosterism, as was its tradition, exaggerated promises. Carey Act developers excelled at promises, and during the first two decades of the law's existence their self-generated publicity had effectively drowned out criticism of the law in Colorado ever since 1895, when outgoing Populist governor Davis H. Waite disparaged it. Except for the most egregious pure promotions, the state land board diligently accommodated developers' plans to irrigate Carey Act lands. Indeed, when the board requested the additional million acres of federal land, it ignored the law's limited impact in Colorado and argued its seeming promise for the arid West: "The Carey Act has been of immense benefit to all of the arid States in settling up arid vacant Government lands that cannot be settled up in any other way." Moreover, the board concluded that the act was increasing population, land valuation, farm production, and taxation for states, counties, and schools—all with proper protection for the small farmer.[3]

Granted its wish, in autumn of 1911 Colorado began accepting applications for new Carey Act projects without concern for exceeding acreage limits—just as the irrigation bond market collapsed and just after the state land board had undergone its major restructuring. The last significant changes to the state land board had been made in 1887, when the legislature created the Office of Register to manage business on behalf of the board—which consisted of the governor, the secretary of state, the superintendent of public education, and the attorney general. In 1910 Colorado voters approved a referred measure from the legislature to amend the state constitution to provide that the board consist of a president, a register, and the new position of civil engineer—all appointed by the governor, with the consent of the state senate, and whose six-year terms were staggered. Left unaltered was the board's administration of the Carey Act and its principal constitutional charge: management of state trust lands to fund schools, public improvements, and other entities.[4]

The constitutional changes to the board in 1910 placed the board's greater duties with the Office of Register. Each commissioner's annual salary was set at $3,000, to be paid from "board income" that was exclusive of monies generated from the lease or sale of state land to benefit schools and other funds. Of Colorado's original 4.35 million acres of federally granted land in 1876, the board had sold 1.14 million acres by 1910. The total of sales receipts from the acreage had exceeded $7.5 million, yet the board's operating expenses could only come from a general appropriation by the legislature and miscellaneous fees. These commissioners would accept the reality that their salaries might be tenuous. In January 1911, Governor John F. Shafroth began his second term by appointing the commission, which the state senate later confirmed: President Edward Keating, whose term was to expire in 1917; Register Benjamin L. Jefferson, (reappointed) term to expire in 1915; and Engineer Blair Burwell to the newly created position of civil engineer, term to expire in 1913.[5]

The restructuring of the state land board was one of many reforms in Colorado during the Progressive era, the political movement that enacted civil service laws, citizen-approved constitutional amendments, and dozens of other direct popular participatory initiatives. In this instance, state legislators—mostly Democrats—seized on allegations that the state land board was rife with corruption and cronyism. In 1909 the solons had called for the investigation of Mark G. Woodruff, a former register charged with criminal fraud and other improprieties in handling state land sales in the San Luis Valley and with a coal lease in Routt County. Since 1887, following every gubernatorial election—which then were biennial—a different person had served as register, the board's manager of daily affairs. Former register Woodruff was cleared of wrongdoing, but charges of fraud were frequent.[6]

Notwithstanding the installment of a newly structured state land board, the public policy of developing Colorado-owned and Carey Act lands was the codification of a public lands policy begun in early 1909. Governor Shafroth, days into his first term, began aggressively selling state-owned lands for three purposes: locating settlers to the lands, bolstering the various state trust funds, and promoting economic development across the state. Shafroth, who earlier had represented Colorado in the United States House of Representatives, had repeatedly called for federal land cession to the states, and his attitude regarding developing unoccupied federal land had not changed. The fact that Carey Act promoters proposed nineteen projects during Shafroth's first term and that the state engineer had received more than 2,000 general filings to divert floodwaters from already severely over-appropriated streams illustrates the extent to which the development of arid lands had become a public policy priority by 1910. Coinciding with the state effort to induce settlement was the national policy implemented in 1909 that extended the Homestead Act to permit a person to file for as much as a half section (320 acres).[7]

The register of the state land board, Doctor Benjamin L. Jefferson, had been appointed register to the previous board by Governor Shafroth in 1909 and was a former state senator from northwestern Colorado. Born in Georgia in 1871, he attended private schools there and went on to graduate from the University of Maryland School of Medicine with degrees in dentistry and surgery. In 1892 he moved to Colorado, settling in Littleton before moving to Hayden in 1895, where he became the town's first resident physician. The area's residents elected him to the state house to serve from 1898 to 1900 and to the state senate from 1900 to 1908. It would have been difficult to find someone more familiar with the challenges of settling people in the state's less populated regions.[8]

Doctor Jefferson's tenure as register of the state land board was remarkable. He had no significant knowledge about public land policy; however, he shared Governor Shafroth's belief in accelerating the development of arid lands across the state. Jefferson hired two extremely capable assistants: Lucy E. Peabody (best known for her efforts to preserve Mesa Verde as a national park) as deputy register and Catherine B. Van Deusen (a writer and editor from Canon City) as chief clerk. Peabody, who would serve the board for a decade, oversaw an office of mostly men—appraisers, surveyors, and wardens. Her influence is clearly visible in preserved records of the state land board.[9]

Jefferson estimated that in 1909, Colorado's vast estate of 3.69 million acres was worth between $40 million and $50 million. Moreover, he believed the properties were a sacred trust, a heritage to the state's schoolchildren. His management of those lands—287,000 acres of which he sold to settlers and developers during 1909 and 1910—boosted trust fund balances, as did revenues from grazing leases, mining and timber sales, and other sources. In addition, his agency provided for the temporary segregation of more than twenty Carey Act proposals that envisioned developing hundreds of thousands of acres, which further shows his office's output. In 1911, on the occasion of the board's reorganization, Governor Shafroth's reappointment of Jefferson as register underscored the state's emphasis on attempting to develop arid lands.[10]

Register Jefferson was a hardworking bureaucrat. He doubled the number of board meetings concerning Carey Act matters, land sales, grazing leases, timber sales, coal leases, and other general administrative issues. On one occasion the governor and the attorney general scolded him for depositing money the land board had collected in banks he favored before he moved it to the state treasurer. That misjudgment aside, Jefferson regarded his charge involving state lands and the Carey Act as vital to Colorado's future. He saw the Carey Act as a logical application of public policy regarding desert lands across the state. The unsettled arid federal lands were no different than the dry, unoccupied state-owned lands. He expressed that view in his agency's 1909–1910 biennial report, in which he stated that such "land should be sold in order that it might derive equal benefits from the fast decreasing opportunity for irrigation, as irrigation regulates and governs all land values." Similar to the sale of state land to home seekers and irrigation developers that allowed land to be bought on installment (up to eighteen years), Carey Act projects afforded a similar opportunity to purchase land (and water) with deferred payments and similar interest terms: 6 percent and 7 percent.[11]

But try as he might, Jefferson and his fellow commissioners could no more make the settlement of state-owned arid lands successful than they could wish a Carey Act project to fruition. By late 1912, Colorado's economy was in the second year of a minor recession that lasted until 1914. Assumptions by policymakers about developing desert lands in Colorado had begun to weaken. The dramatic increase in the sale of state lands in 1909–1910 fell by two-thirds in 1911–1912, to a mere 80,000 acres, depleting the board's operating fund that paid commissioners' salaries. The board's incoming receipts for its obligation funds held steady at roughly $1.5 million on leases through the end of the 1913–1914 period. However, the financial situation of its operating fund showed no indication of improving when only 91,000 acres sold, continuing to reduce collections of transaction fees.[12]

Meanwhile, the irrigation bond market for new reclamation projects in the West showed no signs of recovery. The Carey Act in Colorado had just one completed project: the development at Two Buttes. As early as 1911, the first serious criticism in the state about the law's failure to spur growth came from residents in northwestern Colorado. They complained that the Little Snake and Great Northern Projects were impeding growth by keeping hundreds of thousands of acres from general settlement. In August the editor of the *Routt County Republican* in Hayden made a direct jab at state officials when he wrote of people's aggravation: "Shall the state land board encourage land speculation by giving unlimited time to these companies? Shall the land board hold back the development of this rich country by allowing this monopoly of all public land to continue? Hundreds of settlers are turned away from here because the land is all tied up by the big (at least big on paper) irrigation companies. . . . Make them do something, land board, make them fish or cut bait."[13]

By autumn of 1912, the critical situation of Carey Act projects across the arid West had generated such nationwide protest that the United States Congress asked Secretary of the Interior Walter Lowrie Fisher to investigate the practical operation, history, and condition of all projects in the various states. Fisher appointed three experienced, highly competent federal officials—F. R. Dudley of the GLO, F. W. Hanna of the United States Reclamation Service, and Herman Stabler of the United States Geological Survey—to compile data from their field offices. In Colorado, GLO engineer Charles W. Wells filed reports on specific projects. His comments about the schemes in the state's southeastern corner emphasized the serious weakness of irrigation financing and the reality that a severe drought had impeded development, a point he underscored in his separate report (mentioned in chapter 3) on the Two Buttes Project. Regarding the Valley Investment Company's project on Muddy Creek in Bent and Las Animas Counties, he asserted its viability and noted that "the promoters are such men as reasonably should have been able to have secured sufficient backing to go ahead with the work; but by the time the segregation was approved by the [Interior] Department, irrigation securities were on the toboggin [*sic*] slide downward." Nevertheless, he concluded: "The project is a good one and it should attract the attention of investors."[14]

In February 1913 the special committee presented to Congress a frank and critical report documenting that land patented to the arid states under the Carey Act amounted to 473,999 acres, a fraction of the millions of acres Congress had made available to the states for reclamation. The committee disclaimed any moral responsibility on the part of federal and most state officials for the failure of any Carey Act projects. It offered that no government directly supervised or controlled any of the projects but acknowledged that federal and state officials had not prepared well to administer the law during its initial stages. However, the committee reported that reforms begun in 1909 by the federal and some state governments (those amenable to federal feasibility investigations, such as Colorado) indicated "a narrowing of the field of Carey Act promotions to feasible projects and a very gradual placing of Carey Act operations on a sound financial basis. A more reasonable relation between the area of lands segregated and lands reclaimed is to be expected."[15]

The committee delivered a particularly scathing review of the law's application in Colorado. It noted that companies had sold merely 34,409 acres with appurtenant water rights, but actual land irrigated amounted to less than 640 acres: one square mile. No acreage had passed to the state by patent, so no settler held title to any land. The only completed project, Two Buttes, had not had enough water to deliver to the land even though its reservoir had been constructed three years earlier. The committee judged that the excessive cost of water, such as Valley Investment Company's $60 per acre, and the occurrence of water shortages during lesser than assumed runoff times seriously plagued projects statewide. Colorado's poor results were due, the report concluded, to irresponsible promoters, insufficient financing, insufficient and incompetent engineering advice on water supply, and the chaotic condition of water rights in the state resulting from the lack of administrative control over water appropriations and adjudications.[16]

In response to the report's searing critique of Colorado projects, state land board register Benjamin L. Jefferson traveled to Washington, DC, in April 1913. He met with the state's congressional delegation and officials of the Department of the Interior, but no substantive remedies came from the meetings. Meanwhile, the state land board's operational finances had reached crisis mode in late 1912. Fees collected for the commissioners' cash fund no longer met the needs of office operations and also were unable to fund the salaries of the three commissioners. In addition, the general assembly had made no further appropriations to fund the board. Jefferson's pleas for additional staff and funding fell on unsympathetic ears given a downturn in the state's economy. Newly elected governor Elias Ammons, a rabid states' rights Democrat and fierce anticonservationist, took office in January 1913 and pledged austerity. Board president Edward Keating resigned in February. Jefferson resigned in July after four years on the board and two years before his term as register expired to become President Woodrow Wilson's minister to Nicaragua, where he served until 1921.[17]

Following the resignations of Keating and Jefferson, engineer Burwell resigned. Governor Ammons kept his austerity pledge. The only funded office was that of register; the other members served with no remuneration. Governor Ammons and two succeeding governors followed a practice of appointing the attorney general to serve as board president and the state engineer as board engineer. Notwithstanding the agency's funding issues, the office force persisted in its duties. Lucy Peabody, very much a symbol of the Progressive era, effectively oversaw a reformed agency, changed with good intentions but then neglected by policymakers. (In 1915 the engineer member of the board again began receiving pay; the board president was not paid until 1917, when the commissioners' cash fund balance grew as the agency began to collect fees on increased land sales, leases on grazing, and oil and gas operations.)[18]

To replace Doctor Jefferson as register, Governor Ammons appointed Volney T. Hoggatt—the most flamboyant and opinionated register to serve as a state land board commissioner and a critic of the Carey Act. The state senate confirmed Hoggatt as register in the late summer of 1913. He was a longtime confidant of Frederick Bonfils, one of the owners of the *Denver Post* who was a backer of the Williams Highline (Leach) Carey Act Project south of Hayden. A flamboyant character and a forthright anti-Semite, Hoggatt was born in Iowa and had been a childhood chum of Billy Sunday (later of evangelical fame). He received a degree in agriculture from Iowa State College before gaining acceptance to the bar and establishing a law practice in Dakota Territory, where he helped land-hungry farmers secure homesteads. He later followed the hordes to Oklahoma for the opening of the Cherokee Strip and then went to Alaska during its gold rush before locating to Colorado.[19]

Hoggatt continued his predecessor's aggressive attempt to sell Colorado state land to settlers. But unlike Jefferson, he believed the state should sell *every acre* of state-owned land to settlers for the benefit of the various land trust funds. Moreover, he meant to address the pressing controversy concerning the state's languishing Carey Act projects, an issue that also concerned Governor Ammons. Hoggatt, though himself something of a noted land speculator, believed as did critics of the act that many companies were holding segregations for merely speculative purposes. By the end of August 1913, two weeks after Hoggatt had taken office, he called for citizen comment on the act and notified every company with Carey Act segregations to prepare to report to the state land board. He stated that if companies had made no effort to improve the land, the board would cancel segregation orders and return the land to the GLO for entry to homesteaders.[20]

Volney Hoggatt excelled at bombast, yet his hearings on the state's Carey Act projects were diligent proceedings. Over a two-week period in March 1914, he and John E. Field, who attended as state land board engineer, met with the heads of the remaining fourteen Carey Act companies. Each company's contract with the state land board allowed for administrative review and cancelation with cause. But the board provided that if it canceled a proposal, within a period of sixty days other parties might take over and complete a project. Much was on the line for every company. Carey Act developers pled their cases in state land board's cramped offices in the state capitol building. The principals of the Routt County

Development Company whose Little Snake River failure had become legendary were first on the docket. They offered no defense. Other developers, however, presented cogent arguments that their projects should continue in hopes that the irrigation bond market might recover.[21]

Meanwhile, State Engineer John E. Field, who oversaw the state's water administration, assumed the dual role of state land board engineer. Field was a Denver native who earned an engineering degree in 1888 from the Sheffield Scientific School at Yale University. He had been Colorado's state engineer from 1897 to 1899. Field acknowledged that the present moment required a serious reevaluation of the effort to bring irrigation to the desert lands above the established canals. Field was not openly hostile to the Carey Act's purpose. In private practice with his Sheffield upper classmates Abraham Lincoln Fellows and Michael Creed Hinderlider, he had designed Two Buttes Reservoir and its canal system. The reality that Colorado's dozens of other Carey Act promotions and projects were scattered across the state and none of them functioned was a circumstance the board could no longer ignore.[22]

Attorney General Fred Farrar also served as president of the State Board of Land Commissioners of Colorado from 1913 to 1916. He, like Field, was a Colorado native and very knowledgeable about water issues. His role on the state land board, however, was minimal, and he did not regularly attend its meetings. Both he and Field deferred to the register concerning the agency's policy matters.

By October 1913, Hoggatt's notification to Carey Act developers to show cause for continued state authorization had resulted in the voluntary cancelation of two applications: the Blue Mountain Canal Company (a pure speculation of 177,000 acres in Rio Blanco and Routt Counties) and the Tyrone Canal Company (a 25,000-acre undertaking in eastern Las Animas County bound up in litigation with other water users). Both companies agreed to abandon their projects, and the corporations dissolved. The state land board later notified the GLO, which opened the land to general entry.[23]

Hoggatt's call for citizen comment drew a pointed letter from Farrington R. Carpenter, an attorney from Hayden who would later represent Great Northern Carey Act settlers in their effort to secure individual title to their parcels. Carpenter argued in a letter to the *Routt County Republican* that the Carey Act in northwestern Colorado had permitted the withdrawal of about 600 square miles from legitimate development by hundreds of home seekers. Moreover, those who wanted to expand their ranches and farms were unable to do so because Carey Act companies often controlled parcels that adjoined theirs. The Carey Act projects had never passed beyond the promotion stage, and their segregations were a serious, if not almost fatal, barrier to a homesteader's means of livelihood. Carpenter concluded that at a minimum, the United States Congress should amend the Carey Act to remedy its impediment to development.[24]

One such company pleading that its proposal was viable was the Badito Project, Carey Act Segregation List No. 22. Its developers located the scheme on the upper Huerfano River in the Wet Mountain Valley of south-central Colorado. It was the usual Carey Act drama and one familiar to John E. Field. The Powell Survey of 1889–1891 had located dam Site No. 29 near the old village of Badito (little ford). Its development as a Carey Act project was the idea of Boulder and Denver resident Alonzo P. Sickman, who proposed a 40,000-acre-foot flood storage reservoir on the Huerfano in 1905. In addition to developing an irrigation system, Sickman had hoped to develop a hydroelectric plant to supply the region's burgeoning coalmining economy. The company surveyed two major canals in 1908, one to extend east across Huerfano County and the second to run northeast into Pueblo County. Sickman had persuaded Pennsylvania capitalists E. J. Lesser, Joseph H. Blair, and I. N. Blair to join him in the scheme. The men capitalized the company at the promotional figure of $8 million. In 1910 the company principals renamed the enterprise the Huerfano Valley Irrigation Company and reduced its stated capitalization to $2 million. They applied to the state land board for the segregation of 70,000 acres.[25]

The company hired the Denver firm Crocker and Ketchum to engineer the project. Milo Smith Ketchum, who had taken a sabbatical as dean of civil engineering at the University of Colorado, redeveloped earlier designs and concluded that water supply from the Huerfano's upper drainage area of more than 500 square miles was sufficient to irrigate the stated 70,000 acres. In October 1910 a skeptical state land board sent the redesign to then state engineer Charles W. Comstock, who found serious fault with Ketchum's conclusions. Comstock recommended the segregation of only 20,000 acres, citing over-appropriation of water rights on the

Hurefano, the river's erratic flow, and the "fact that it [southeastern Colorado] is the most difficult district in the state in which properly to distribute water."[26]

Ketchum later satisfied Comstock's concerns by proposing a second flood storage impoundment to capture surplus water from the Greenhorn and Graneros drainages near Rye, south of Pueblo. The state land board then allowed the Huerfano Valley Irrigation Company a segregation of 36,779 acres. The GLO did not block the revised segregation, and the state land board granted it during the summer of 1911—the moment when the irrigation bond market had collapsed. Nevertheless, the company contracted with the state to develop the project and pressed on with attempting to finance it, which the company estimated would cost $1.49 million.[27]

Huerfano Valley Irrigation Company hired Havermeyer Construction Company to build the system. However, Havermeyer demanded an audit and hired Winfred L. Rucker, a respected Denver bond broker, and John Philip Donovan, an engineer, to assure the project's feasibility and that it did not conflict with the nearby development of the Pueblo–Rocky Ford Irrigation Company. The latter company had interest in the Orlando Reservoir northeast of Walsenburg and hoped one day to supply water to sugar beet farmers. In addition to providing data for the construction company, the Rucker file was a conveniently packaged document for potential investors, presuming the irrigation bond market rebounded. Whether the file ever made its way to a bond house is unknown. In early 1913 the project affiliated with the Colorado Southern Irrigation Company, which operated the large De Weese system between Cañon City and Pueblo and in the Wet Mountain Valley.[28]

The Badito Project represented the kind of conundrum the state land board faced. The company had exercised due diligence in advancing the proposal but had reclaimed no acreage. The state land board chose not to cancel the project. But the project never did issue bonds. And as permitted by the Carey Act, it exercised the full advantage of its ten-year time extension period for completion, although construction never occurred. The delay frustrated residents until 1924, when the GLO opened the acreage to general settlement after a year-long study by the United States Reclamation Service that found water rights had been decreed in excess of both normal and flood-flow levels, and the construction cost would be prohibitive.[29]

Fred L. Harris of Two Buttes Irrigation and Reservoir Company presented a report to the board on March 18, 1914. The report satisfied the board, despite a board appraiser's earlier conclusion that the project was one of the biggest frauds in Colorado history. Meanwhile, unusually heavy snows were melting in Baca County, and company representatives reported that they were optimistic about the reservoir filling.[30]

The officers of Valley Investment Company met with the board on March 24, the last day of hearings. Dr. Alpheus Pollard and Frank Whiting could offer little evidence of their progress, except for about $40,000 spent on surveys and the construction of no more than three miles of the Smith Canyon Canal feeder to the planned Muddy Creek system. They presented no financing plans. Hoggatt, as he did with each company, pressed the issue of whether the company was simply a speculation—a scheme holding onto land and keeping it vacant to prevent other settlers from locating there. The company officers gave no response.[31]

In early May 1914 the state land board canceled five Carey Act segregations. The *Weekly Ignacio* (CO) *Chieftain* reported that the board acted on the grounds that "laws have not been complied with either in the proper pushing or completion of the projects": the Toltec Canal Company (16,000 acres in the San Luis Valley), Pueblo and Northeastern Irrigation Company (15,280 acres in Pueblo County), Dolores Irrigation Company (196,650 acres in Montrose, Dolores, and Montezuma Counties), Routt County [Little Snake] Development Company (39,000 acres in Routt and Moffat Counties), and the Valley Investment Company (24,000 acres in Bent County). When included with the two earlier voluntary cancelations, the total was almost 500,000 acres. Register Hoggatt sent immediate notice of the board's action to the GLO.[32]

Hoggatt summed up the results of the hearings in the state land board's 1913–1914 biennial report and made clear that they demonstrated the "absolute failure of numerous segregations to make good." He argued that the State of Colorado had a minimum obligation to reimburse settlers for the cost of the land (the twenty-five cent initial payment) on those developments that had failed to undertake any construction. He was most incensed with the Routt County Development Company's lack of development on the Little Snake River. He did not mention the company by name but fumed that it had issued $400,000 in bonds "based on, Lord only knows what, and these bonds were sold to the innocent purchasers, with the promise that the amount realized would go into the construction of canals and reservoirs. I do not believe that over $50,000 was ever efficiently used in construction; the balance has been frittered away, the company is bankrupt, and the state has no one to blame but itself for permitting this condition of affairs to exist."[33]

The well-publicized Carey Act hearings, in addition to culling half a million acres from state oversight, resulted in the land's return to the United States Department of the Interior for general entry. Developers did not add 1 acre of the additional 1 million acres the United States had granted Colorado in 1911 to the state's Carey Act designations. Register Hoggatt called for the state to assume responsibility for the act's failure by appropriating funds for the completion of some projects and reimbursing individuals who paid the twenty-five- or fifty-cent cost for the land.[34] Both recommendations were gratuitous. The general assembly ignored them. Moreover, state law constitutionally restricted the use of state land board obligation funds. And its commissioners' cash fund, from which any reimbursement might have come, had become so depleted that two commissioners were serving without pay.

Left standing at the end of 1914 were only ten Carey Act projects in Colorado that totaled no more than 550,000 acres. The state land board's hearings emphasized the importance of canceling projects but also highlighted the companies' serious financial condition. Only the Two Buttes Project offered promise. Developing the state's remote benchlands by reclamation, be it Carey Act projects or other large-scale private schemes, was reaching a breaking point. Across the American West, similar patterns of failure of promotions, unstable financing, and seemingly untenable circumstances cast serious doubt about such development in general and the Carey Act in particular. The missteps were many. Ironically, three weeks after the Colorado state land board ended its Carey Act cancelation hearings, western governors and the nation's most prominent reclamation experts gathered in Denver to address the ongoing crisis in Carey Act reclamation development. There, two voices from the past—from the genesis of the Carey Act—captured the moment and the ongoing conundrum of reclaiming arid lands.

6

PERHAPS NOBODY IS TO BLAME

Joseph M. Carey, still at war with nature's limits, departed from Cheyenne to Denver during the first week of April 1914. Two decades had passed since he traveled to Washington, DC, with plans to spur private irrigation development on arid lands. He was now age sixty-nine and governor of Wyoming. His purpose in Denver was to chair the Conference of Western Governors, the group's third annual gathering to discuss various aspects of the relationship between western states and the federal government's control over much of the region's land and natural resources. The governors met in the ornate chambers of the Colorado State Senate, just down the hall from the offices of the state land board where two weeks earlier the board had concluded its special cancelation hearings on Carey Act projects. Following the governors' conference, a general irrigation conference including the governors, federal officials, and prominent reclamation advocates took over the state house chambers[1]

Governor Carey, one of the grand old men of Wyoming lore, had not prepared well for the meeting. But he was an irrepressible personality with a

https://doi.org/10.5876/9781646426492.c006

command of public policy issues equal to that of any of his fellow governors. He had been elected Wyoming's governor in 1910 as a Democrat, his first major office since 1895 when his state's legislature refused to return him, a Republican, to the United States Senate. He was banished from that federal office largely because he had not supported the silver cause. Carey, now a Progressive, became chair of the Denver gathering at the conclusion of the previous Conference of Western Governors in Salt Lake City in June 1913.[2]

At the Salt Lake City gathering, Governor Carey unleashed a verbal counterattack on the harsh criticism congressional and state authorities had leveled against his only notable congressional accomplishment. In a speech titled "What the Carey Act Has Done for the West," he declared that "the law had experienced no more mishaps than are generally incident to the undertaking of great projects—no more troubles than have been encountered by private organizations in railroad building, etc." He argued that the Carey Act had worked in Wyoming, where there were twenty-five completed projects on thirty-nine segregations totaling 297,499 acres. Developers had built the projects at an estimated cost of $4.7 million. His figures, however, were entirely his, an argumentative construction because Wyoming had refused to cooperate with the federal investigation of Carey Act projects in 1912. To his audience in Salt Lake City, he conceded that the fact that "mistakes have been made cannot be denied. There has been carelessness, and money had not always been made by those who promoted the scheme. This is not the fault of the law but rather due to lack of experience, want of capable engineers so far as plans and specifications are concerned, and lack of good management in the protection of the companies from very large and unnecessary expenditures of money, which have often amounted to actual waste." He surmised nonetheless that the law had the potential to "redeem" fully 20 million acres in the desert land states. He could not bring himself to accept the overwhelming evidence that Carey Act developers across the West had embraced a failed business model.[3]

At the Denver conference in 1914, Governor Carey had not planned to discuss the Carey Act, even though that act, the leasing of public lands, and the idea of using regional banks for financing development were the meeting's principal agenda items. Not only were Carey Act projects failing by the dozens across the West and the law proving ineffective in reclaiming great swaths of arid lands, also failing were many new irrigation district projects. The lack of investor interest in irrigation construction bonds combined with water supply shortages and too few settlers with sufficient ability to generate revenues made the arid lands issue a concerning topic for every western governor. Economic and community growth were chief goals of each administration, and the commodification of land and natural resources played an essential role in those efforts. Moreover, United States Reclamation Service projects continued to experience major financial shortfalls, in part because of systemic problems with affording settlers profitability on many of the service's numerous projects. The nine attending governors were eager to express their concerns to United States Department of the Interior officials who had traveled from Washington, DC. Among them were Andrieus A. Jones, first assistant secretary; Clay Tallman, commissioner of the General Land Office (GLO); and Frederick Haynes Newell, director of the Reclamation Service ever since its creation in 1902. The purpose of the irrigation conference to follow was to examine in further detail ways and means of making new irrigation development more sound.[4]

Governor Ammons welcomed his guests and asked "how can we help the Carey Act?" Remarkably, he made no mention of the law's abysmal failure in Colorado, and he would make no reference to it throughout the conference. An anti-conservationist zealot of the first order, he had nearly hyperventilated at the Salt Lake City conference, ranting at length about the evils of the federal government owning and controlling one third of Colorado and much of the West. His attitude at the Denver gathering was equally hostile regarding federal conservation of national forests and federal preservation of national parks and monuments. The federal government, he said, could "very well afford to give a helping hand" to assist development in the western states. With that, he asked Governor Carey to speak.[5]

Carey took the floor and extemporaneously reiterated the general theme of his Salt Lake City address: that the law's only fault had been in its administration. As if to validate the old argument of western politicians, such as one who had advocated for the cession of arid federal lands to the western states as necessary for stimulating economic growth, he asserted that Wyoming had reclaimed 900,000 acres under the Desert

Land Act to which the Carey Act was appended. Governor Carey framed his remarks this time in broad philosophical terms that tapped into the deeply held beliefs of western individualism, agrarian values, and the general disdain of federal authority felt by the executives in attendance. "The development of this western country," he said, "from the defeat of Braddock, near Pittsburgh, up to the present time, had been the result, more or less, of accident. The disposition of men who came to this continent has been to better their conditions; to get farther out; to have elbow room; to have a chance in life." The growth of the West from the time of the French and Indian War during 1755, he continued, occurred "because the whole theory—everything—was based upon this one thing: the more widely we parcel this land out among all the people, the greater the inducement we afford them to farm, to get holdings, to find mineral[s], the more rapidly will these territories settle up. Settling up they become additional states of the Union, they add to the glory of this country in peace, and will materially add to the strength of this country if we are so unfortunate as to get into a condition of war." Federal control of all these things, he said, would bring about stagnation in the western states that was never experienced by states in the Midwest. A federally forced plan was contrary to the western experience, and he placed major blame for administrative failures of the Carey Act with the Reclamation Service, which, he asserted, was the law's only enemy and jealous of power. Nonetheless, he believed a united West would bring about a prosperous West. He concluded that "irrigation is not at first a financial success—I do not care under what conditions you commence it."[6]

No governor challenged the basis of his argument for want of appearing divided on the policy of growth in the West. Governor Tasker Oddie of Nevada worried about the general loss of settlers to subsidized Canadian farms. He said his state had experienced Carey Act frauds but had reformed its law to be more parallel to those of Wyoming and Washington. Governor John M. Haines of Idaho, a real estate developer, acknowledged that his state had utilized the Carey Act more than any other state: "I can say for myself—doubtless as well for them [other members of his state's land board]—that we have our troubles. We have all kinds of projects—good, bad, and indifferent; and it has really been the effort of my life in the past year to study all of these projects, and this work, and this development." Idaho's difficulties with the law, he said, were different from those in the other states because colonists had arrived too soon, before developers had undertaken many of the projects. Nonetheless, he emphasized that development was occurring on millions of acres of land, and assessments were in the millions of dollars.[7]

Goernor Oswald West of Oregon tried unsuccessfully to refocus the discussion on failed Carey Act projects and to elicit a comment from Governor Ammons, who was uncharacteristically mute. West, a onetime state land agent, mentioned Colorado's abysmal record of having reclaimed just 640 acres of the tens of thousands of acres of arid lands it had segregated. He apparently was referencing early reports of the Two Buttes development He continued: "The conditions in Oregon have been bad." Nearly every project "has been a failure; and it is the problem of devising some means to put these projects on their feet, and to save the settlers the money they are about to lose. . . . We in Oregon are so deeply interested in that question . . . that I feel it my duty not to let this conference get away from the main question."[8]

The conversation drifted back to perceptions of the law's administration when Governor William Spry of Utah acknowledged the fraudulent role of speculators in certain projects and the weaknesses with financing projects. But he pointed to the Delta Project, in southwestern Utah, as one of his state's fine examples of the law's beneficial application. The law, he asserted, "has been an experiment, perhaps; nobody is to blame; I am not criticizing anyone, because I do not think there is any criticism due . . . but these things have fallen down, and for that reason the Carey Act has, unfortunately, come into disrepute. It is unfair to the act; it is unfair to the mover of the act, because the Carey Act means eventually, if properly carried out, the reclamation of the entire West."[9]

Assistant Secretary of the Interior Andrieus Jones, a New Mexican, listened as his fellow westerners unleashed their frustrations about the plight of new reclamation projects in general. At the irrigation conference that followed the governor's conference, more than 300 attendees from various western states filled the house chambers. Perhaps the chief executives believed President Woodrow Wilson's appointment of Department of the Interior officials from the West, such as Jones and Tallman of Nevada, were not as wedded to the nationalism of conservation as were previous officials in the department during the Roosevelt and Taft administrations. Secretary of the Interior Franklin Lane, a Californian, was not in attendance, but his earlier authorization of the Hetch Hechy Reservoir in Yosemite National Park suggested that states' development interests regarding public lands might have greater standing than had been the case in previous years.

Jones had hoped to hear from the governors and the irrigation conference attendees about specific proposals the federal government might undertake to help rehabilitate irrigation development in the West. Hearing a litany of grievances instead, he offered some of the department's proposals regarding existing projects. For Desert Land Act and irrigation district schemes, he called for more thorough feasibility investigations and coordinated efforts by state and federal agencies so settlers could obtain capital at the lowest interest rate. In the case of Reclamation Service projects, such as the Uncompahgre scheme in Colorado where the service had hiked the cost of water to settlers, he proposed loaning the settlers money to improve the land and extending the terms of repayment from ten years to thirty years. He asserted that the policy changes were possible because the federal reclamation fund was $100 million, and the United States Congress could authorize issuing additional bonds with interest payments guaranteed by the federal government.[10]

For feasible Carey Act projects, the assistant secretary suggested providing "a fund to start and finish a project; put the farmer on the land; give him water; look after his wants and earnings; and turn the whole thing over to a local association." Thus, with a project completed, perhaps liens could be made so a settler might be able to dispose of the bonds to repay the newly created fund. Jones made no suggestion for dealing with the problem of making the land, which was still federal land, collateral for bonds. Perhaps Congress, he offered, could be persuaded to use the Reclamation Fund for defaults on interest payments. In that way, a 3 percent or 4 percent interest cost, instead of 6 percent, might be possible with the federal government practically standing behind the bonds. To what extent he believed such policy changes were possible is uncertain, but his suggestions, unlike the ramblings of the governors, were at least constructive.[11]

Beyond the spotlight shined on the status of troubled reclamation projects in the West, the 1914 gathering of western governors and reclamation specialists in Denver might have had little lasting relevance had it not been for the remarks of an attendee in the delegation from California. Elwood Mead was fifty-six. More than three decades earlier, while on summer leave from teaching in Fort Collins at Colorado Agricultural College, he had slogged through the irrigation ditches of northern Colorado mapping systems for State Engineer Edwin S. Nettleton. Mead, who had been Wyoming's first state engineer and one of the co-drafters of the Carey Act's specific regulations, had long since come to terms with the 1894 law's ineffectiveness. By 1897, Wyoming's Carey Act projects had stalled and applications for new projects had ceased. The cost of building ditches and attracting settlers, he wrote, seemed to be greater than the value of the land and water. Both Mead and US senator Francis E. Warren came to believe that the law's defects were unrepairable, holding that only more liberalized federal land cession laws would be the most effective path to reclaiming arid lands.[12]

Mead had opposed the Reclamation Act of 1902, favoring state- and settler-centric development of the arid lands. Yet he did not entirely reject a national reclamation purpose when he headed the United States Department of Agriculture's Irrigation Investigations Office from 1899 to 1907. His tenure there entailed developing cooperative programs with state engineers and state experiment station personnel to reform water laws and acquaint farmers with irrigation methods. He also taught part-time at the University of California at Berkeley and wrote extensively about irrigation institutions, emphasizing the social and political impact of water development on arid lands. Frustration with his federal position and the university's finances coincided with an offer from officials in the Australian state of Victoria to head their State Rivers and Water Supply Commission. Given wide authority there, he oversaw the development of Victoria's complex streams that comprised much of the Murray River system, the continent's only important watershed. His work included not only engineering waterworks but also settling water disputes with the states of New South Wales and South Australia, as well as colonizing the reclaimed land.[13]

Mead's visit to Denver in 1914 came during the seventh of his eight years in Australia. He was in the United States for personal reasons and to promote his unique Australian farm settlement program. The Denver gathering provided an ideal opportunity for him to reclaim his relevancy in US reclamation policy circles by explaining how the Victorian experience of "a national policy of aid to settlers on irrigated lands will prove of immense value to the development of this country and stop the drift of American farmers to other [foreign] lands."[14]

Surveying the crowd, the engineer must have noticed old friends and associates; he also would have seen his old nemesis, Frederick Haynes Newell, director of the United States Reclamation Bureau. The director's star, however, was fading from the national scene. By year's end, Arthur Powell Davis, the nephew of John Wesley Powell, would replace Newell. The rift between Mead and Newell dated to the late 1890s when Mead was at the Department of Agriculture. Newell had also walked his share of Colorado's desert lands when he was chief of the hydrographic branch of the United States Geological Survey. Their early years with the separate federal agencies were formative years for them both. Mead's vision of an agrarian society became more defined as he came to specifically focus on the farmer and the complexities of western water law and desert agriculture, which he always maintained necessitated local state control but different in shape than the Carey Act. Newell, meanwhile, focused on hydraulic works and came to advocate for a federal reclamation program. The ascension of Theodore Roosevelt to the presidency in 1901 elevated reclamation in the national discourse. Indeed, Roosevelt specifically sought ought Newell and Mead (as well as Gifford Pinchot) to make suggestions for his inaugural address—ideas the president used to call for national reclamation but with the distribution of water left to the settlers under state laws. In the complex debate that raged over a centralized national reclamation program versus a decentralized one, Mead had drafted an unsuccessful alternative bill to the eventually successful Newlands bill that became law. Mead's star faded then and Newell's star brightened.[15]

But this was a new day. Elwood Mead titled his address "Systematic Aid to Settlers Is First Need." Very few attendees in the chamber were unfamiliar with Mead's acclaimed work in Australia. He meant to challenge American assumptions: "The weakness of American irrigation development has been its exaltation of works and neglect of the settler who uses and pays for them. Irrigation schemes have failed because the settler has been left to struggle unaided with a task beyond his means and strength."

He said it was time to stop chasing rainbows and deal with realities to aid settlers and lessen the losses and hardships unaided development entailed. The absence of adequate financial help, he underscored, was the main cause of the stagnation in irrigation development: "One only needs to put himself in the place of the settler to realize what a costly and serious venture it is to attempt to transform unimproved land into an irrigation farm. . . . Before a settler can have any return from his land he must do many things. . . . A house must be built, ditches dug, land cleared and graded, seed sown and the somewhat difficult art of irrigation mastered under untried conditions. While this is being done there is no income. His scanty capital is being swallowed up in living expenses." Mead then stated the crux of his larger argument: "To those have been added, in recent years, great increases in charges for land and water. Great dams and costly and permanent works mean much higher water charges than were paid by the earlier generation of irrigators. . . . With water rights costing from $40 to $60 per acre, and with the present western interest rates, the chances are all against the success of the settler who has less than $5,000 or $6,000 capital, and the question which now needs to be decided is whether this nation is to restrict opportunities under national or private works to men with this larger capital, or encourage poorer men by helping them to improve their farms."[16]

"Thus far in America," Mead continued, "we have almost entirely ignored the requirements of colonization and settlement. We have looked upon the building of irrigation works and the marketing of irrigation securities as the problems of irrigation development. . . . Another mistake has been to regard irrigation enterprises as something which could be paid for quickly. We have taken it for granted that if they were once built the farmer would come forward and foot the bills." Mead could just as well have been describing the condition of every Carey Act project in Colorado. He continued: "The actual facts are entirely different. Irrigation works do not create irrigated agriculture. The money spent on dams and canals must be followed by an equal or greater expenditure for houses, farm buildings, fences, grading and ditching fields before the water can be used and irrigation works have either revenue or productive value." He flatly stated that until the existing American reclamation model changed, irrigated land would continue to involve regrettable hardship and loss for deserving settlers. He added that poor men and American farmers would continue to immigrate to the already irrigated farms of Australia and Canada.[17]

Elwood Mead was openly challenging the cherished American ideal of self-reliance—the idea that if something were free, an individual would have no reason to work hard to succeed. Facing governors, federal officials, and other delegates, he described his experience in Australia and how Victoria's state government had restructured its unprofitable irrigation developments. He focused on the state's systematic aid to settlers to address conditions he believed were similar to those that existed in troubled irrigation districts, Carey Act projects, and Reclamation Service developments. The Australian state allowed nearly anyone who had industry and thrift to secure an irrigated farm even if the individual had little or no money. Only a small down payment was necessary. The state also provided adequate aid and direction; no initial charge for water rights and only 4 percent interest to cover the cost of the works and maintenance; houses built for settlers on cash payment of about one fourth the cost, with the remainder paid over twenty years at 5 percent interest; and the state grading and seeding portions of the farm. The state followed up by loaning the settler 60 percent of the cost of improvements. He stated: "I have never seen elsewhere men work as hard or achieve as much in the first two years as in those Victorian settlements."[18]

Nearing the end of his remarks, Mead asked: "The question for consideration by the United States is first, whether it is right to subject the settler to the hardships he has now to endure, even if he were able to survive, and what is more important, whether it can continue development unless it offers equal opportunities with other countries." The United States was, frankly, losing its ability to attract farmers. That said, he still saw the tenant farmers in his native Midwest as the best group of settlers to populate the West's projects. He proposed a financing model like that of Victoria's State Savings Bank, which had deposits of $100 million.[19]

Developing new rural farming communities in Australia by tying community planning to irrigation was rightfully among Elwood Mead's chief accomplishments. His closer settlements, as Australians called rural developments, were frequently benchland locations far from the easily irrigable regions of the Murray River system. Self-sufficient communities of irrigated farms of various sizes, a small nearby town, and cooperative exchanges were at the heart of his developments meant to create colonization opportunities for experienced farmers. Closer settlements were also at the core of his broader irrigation development objectives of hydraulic engineering and interstate cooperation. Stable, government-backed financing for settlers was justified because it might lessen Australia's dependence on imported foodstuffs as well as justify the high cost of building the continent's large dams and canals for irrigation and navigable locks for commerce. Mead directed the reclamation program through Victoria's State Lands Purchase and Settlement Board, which began locating settlers from England and other countries in 1909. He helped establish more than thirty developments. Mead's Australian experience, in a sense, was an extension of his earliest notions for the Carey Act's possibilities without the burdens of conflicting government agencies and free-market capitalism's dependence on the fluctuations of irrigation bond financing.[20]

Mead's address to the Denver conference had to be descriptive. He chose not to speak about the deeper forces that made the Australian experience fundamentally different from reclamation development in the United States, particularly for private enterprises such as Carey Act projects. For all of Australia's resemblance to the American West—its aridity, gold rush, economic depression in the 1890s, bankruptcy of its private irrigation enterprises, and other similarities—the role of "state socialism" set it apart. Not to be confused with global labor socialism, state socialism's infusion into Australian policymaking became a more prominent theme during the federation movement and the formation of the Commonwealth in 1901. Closer settlement acts passed before and after federation were the embodiment of Australian state socialism. These laws permitted a state government to purchase portions of large, private grazing estates and subdivide them for the benefit of farmers of limited means. Equalizing opportunity for people on the desert lands, its advocates argued, benefited the development of the state, including the landed gentry. Mead's entire corpus of work in Australia intertwined with the concept of state socialism. He reasoned that the system actually advanced democracy because of its tangible benefit of developed irrigated agriculture and its intangible benefit of agrarian values—elements he believed were the basis of democracy.[21]

Elwood Mead, in many ways, was a man of contradictions. Similar to his failure in Denver to acknowledge why the Australian settlements were even possible, he made no mention of the farmers' growing discontent with his grand design. They had begun to face the reality of hard work, poor crop prices, and the consequences of severe droughts. He blithely ignored the complaints. "Pettifogging critics" he called them. He returned to the United States in 1915 and later spearheaded California's effort to revive farm life with his demonstration colonies near Durham and Delhi, which he patterned after his Australian model. When the two colonies failed for reasons similar to those with which the Australian settlements struggled, Mead had moved on to head the United States Reclamation Bureau—ironically, an offer that had been made but that he had turned down while he was in Australia.[22]

During the Denver irrigation convention, Mead surely encountered Governor Carey. What transpired between the two is unknown, but Joseph M. Carey seems never to have publicly acknowledged Mead's role in the Carey Act. Mead, for his part, seems not to have wanted any attention. In his address, Mead purposefully linked Carey Act schemes, private irrigation projects, and Reclamation Service projects to infer that they all had sufficient water supplies for viable development. Such supplies may have been adequate in Idaho and Wyoming, but in Colorado and other states, Carey Act proposals were often conceived in locations where Reclamation Service projects were unviable. In any event, policymakers attacked Mead's address as un-American and dismissed it outright. GLO commissioner Tallman commented that the occasion afforded "a healthy exchange of ideas."[23]

The gathering of dignitaries in Denver proved more performative than substantiative. Laissez-faire economics ruled the day. Except for Elwood Mead, most of the attendees abhorred government interference in the free market but gladly accepted government handouts such as Carey Act lands. The failure to finance private reclamation projects, as was the problem during the 1890s that Senator Carey thought his bill would solve, was the result of too much capital chasing too few productive investments. In short, the lack of private reclamation financing following the irrigation bond market collapse in 1911 was one of the classic examples of boom-and-bust economic cycles common in the nineteenth and twentieth centuries. Like railroads and mining, irrigated agriculture experienced constant booms and busts, and yet it continues.

In the weeks after the governors and the irrigation dignitaries had left Denver and the issue of Carey Act failures to the vagaries of the marketplace, mountain snows across Colorado began to melt. The winter of 1913–1914 had been one of the state's greatest snowfall seasons. By July, reservoirs across the state had risen to historic levels. Two Buttes Reservoir finally began to fill, and the land was blooming. Meanwhile, south of Las Animas, the Muddy Creek and Smith Canyon development appeared to have died, but there too the green prairie grasses showed signs of renewed life.

7

GEORGE E. O'BRIEN AND THE DISASTROUS MUDDY CREEK COLONY

After saddling their horses in the early summer of 1914, hydraulic engineer Frank H. Whiting and fellow Valley Investment Company investor George E. O'Brien rode across portions of the 24,000-acre Carey Act proposal south of Las Animas, which the State Board of Land Commissioners of Colorado had canceled two months earlier. The Heinan Flats, like the entire Purgatoire region and all of southeastern Colorado, were as lush and beautiful as either man had seen them. They had much to discuss, though nothing of the import of the debates at the Conference of Western Governors in Denver that followed the state's cancelation hearings. Meanwhile, Dr. Alpheus L. Pollard, who had long headed the project in hopes of windfall profits, had recently abandoned it.[1]

Boundless, however, is the developer's mind. Whiting and O'Brien resolved to continue their plan to alter nature's limits in the region and to largely bypass the collapsed irrigation financing markets, which had lost confidence in private reclamation starting in 1911. As provided by the state land board, other parties could assume a Carey Act development

https://doi.org/10.5876/9781646426492.c007

FIGURE 7.1. George E. O'Brien, ca. 1917. Author's collection.

plan within sixty days of a cancelation. Whiting and O'Brien notified the board that they wanted to restructure Valley Investment Company as a mutual irrigation company whereby the purchasers of water rights would forgo deferred payment contracts and instead pay a majority cost upfront to reclaim their proposed development. Their self-financing proposal notwithstanding, the venture remained a Carey Act project, and no other canceled project sought to restructure. Except for the Two Buttes enterprise, all other Colorado Carey Act schemes were either dead or headed for extinction.

By midsummer 1914, the state land board had transferred all of Valley Investment Company's assets to the Mutual Carey Ditch and Reservoir

Company, with O'Brien as head of the new enterprise and Whiting its secretary. The board readily endorsed the new company plan that required settlers to pay only the absolute cost of construction, with no promotion expenses. Settlers would self-fund the Muddy Creek Reservoir and its distribution canals (estimated at about $440,000) at a cost of $30 per acre, plus the 50-cent cost of the land. Valley Investment Company's cost had been $60 per acre. As for the supplemental Smith Canyon component (estimated at $324,000), the plan proposed to fund it through financial channels, presumably a bond issue the mutual company would collateralize by giving a deed of trust on the developed Muddy Creek system. The new company arranged to escrow settlers' money with the Interstate Trust Company of Denver. As required by the state, only the land board could order the release of money from the fund. Further, the board held firm to the requirement of Carey Act projects that such escrowed monies were not to be available for construction until, in this instance, the mutual company sold settlers at least 14,000 acres of 24,000 acres of water rights.[2]

Farmers throughout Colorado and indeed across the arid West had used the innovative convention of forming mutual water companies. Indeed, the end stage of a Carey Act development might be a mutual irrigation company after settlers paid off their deferred payments, with each irrigator then owning a proportional share in the company. Such companies were an effective means of managing irrigation systems that eastern US- and European-financed land development companies had built and then divested after they sold the land to individual farmers. The greater portion of all irrigation systems in Colorado in 1914 were mutual companies. The Mutual Carey Ditch and Reservoir Company, although of different origins than other mutual companies but still overseen by the state land board, was to function similarly to all mutual irrigation companies—privately owned stock companies that operated as nonprofit, at-cost enterprises. The formation of a mutual company provided a means to finance construction costs, although more often it funded other functions as well, such as the legal expenses and operational and maintenance costs of a waterworks. Mutual water companies, unlike the state's irrigation districts, have never had the authority to tax, relying instead on compliant debt payment. Frank Whiting, George O'Brien, and the state land board saw in the mutual irrigation company's design of shared ownership a means to advance the colonization scheme. Although settlers' upfront cash payments might cover the cost of constructing the Muddy Creek component, the totality of the project was dependent on a rebound of the irrigation bond market. In this sense, the formation of the Mutual Carey Ditch and Reservoir Company lends support to two modern arguments of environmental historians who see it as either exploitive of nature for profit or a practical response to aridity.[3]

For Whiting and O'Brien personally, the creation of a mutual company meant they had to come to terms with forgoing windfall profits. The major motivation of Carey Act developers was the hope for extraordinary returns, which kept about half a dozen proposals still before the state land board. Neither of the two men left a record of any reconciliation or a reason why they stayed with the project. But profitability had always motivated O'Brien, who had spent more than a decade building his very profitable Denver grocery store. O'Brien, who was born in Denver in 1874, had become enthralled with the rough and tumble pioneering life of his wife's grandfather Philip P. Gomer, who had brought his family across the plains in 1860 and established a successful Colorado lumber business. George O'Brien studied Gomer's inclination to take financial risks, not necessarily careless risks but calculated ones. Perhaps O'Brien also saw in Gomer's adventurous life a path that his father, John E. O'Brien, could not follow; a Civil War injury and tuberculosis limited his work as a shoemaker and his station in life.[4]

In any event, after his father's death in 1909, George O'Brien persuaded his mother, Kate Rock O'Brien, to file a Desert Land Act claim to 320 acres that abutted on the south a parcel of Valley Investment Company's proposed Carey Act project in 1910. Residency was not a requirement of such claims, but taking steps to irrigate the land was. George O'Brien and Charles Stanley, a company investor with an adjacent Desert Land claim, spent more than $2,000 sinking two wells meant to irrigate a small portion of both claims. Like many windmills that settlers in the West had thought they might use to meet the act's irrigation requirement, the men's wells proved unsuccessful. Nonetheless, George O'Brien built a small house and a barn and hired a laborer to dry farm the required acreage necessary to purchase the land under a March 4, 1915, amendment to the Desert Land Act, which permitted obtaining patent to the land. Thus,

George O'Brien had a financial stake in more than simply his Carey Act project, and perhaps that reason factored into his decision to forgo possible windfall profits and continue with the enterprise.[5]

As president of the Mutual Carey Ditch and Reservoir Company, O'Brien also headed its two subsidiary businesses: States Construction Company, which managed building the waterworks, and the Carey Land Sales Company, which sold water contracts and land parcels to settlers. His leadership of the company beginning in 1914 coincided with the start of a booming US economy as the European war demand for American goods, especially agricultural products, rose to unprecedented levels. Throughout World War I, the economic opportunity offered by US agriculture lured people to farming. George O'Brien sold what was still very much an American dream—to own a farm, work the land, reap a bountiful harvest, and pass on the family farm to one's children.[6]

George O'Brien closed his Denver market in 1918 and began selling the Carey Act project. Although no expenditures could be used to promote the project, he eventually self-funded one flyer complete with a map of the segregation, the terms of purchase, and photographs of lush sugar beet fields in the Arkansas Valley. He advertised by word of mouth and in press releases that ran in both in-state and out-of-state newspapers. Purchasers bought individual units of 20–160 acres at $30 per acre for each water share. Two dollars and seventeen cents (7.26 percent) of each share sold went to expenses incurred by the Carey Land Sales Company, which compensated O'Brien. Depending on the acreage a shareholder purchased, the cost ranged from $610 to $4,880.[7]

O'Brien sold water contracts to families on farms and in cities—people with expendable income who were not averse to financial risk. Like others who had purchased Carey Act water contracts across the state's projects, they represented an exclusive, white, middle-class group of men and women. Many were residents of northern Colorado and were connected to the sugar beet industry. The group included a factory foreman, a blacksmith, many farmers, widows, and other women who purchased contracts themselves or in conjunction with their spouses. In the Las Animas area purchasers included two retired physicians, a farm-implement dealer, a minister, a banker, a hardware merchant, and several descendants of pioneers John Wesley and Amache Prowers. An important figure who bought into the scheme was William A. Colt, a major Las Animas contractor who would build the waterworks. Some people may have believed it was an opportunity for their returning doughboy to tap into the booming economic possibilities of irrigated agriculture. George O'Brien sold his son, Gerald G. O'Brien, an acreage that abutted the O'Brien Desert Land Act claim.[8]

By early 1919, George O'Brien had sold nearly 200 water contracts to prospective colonists, even as the Spanish Flu pandemic raged. If the purchasers were skeptical about the project's viability early on, it is not apparent in the historical record. In the meantime, a few investors seeking to enlarge their parcels began filing land entries on free federal lands near the segregation, as O'Brien had done in early 1909. People had their choice of filing under the Desert Land Act, the Homestead Act (1862), the Enlarged Homestead Act (1909), or the Stock-Raising Homestead Act (1916), which offered a parcel up to 640 acres.

Homesteaders filing on federal lands surrounding the company's segregation were, like others in the state, participating in Colorado's last opportunity to prove up free federal land. Indeed, across eastern Colorado, homesteaders had already claimed much of the free land. The belated homesteading boom, however, proved most enticing to home seekers in northwestern Colorado, where vast stretches of government land constituted hundreds of thousands of acres—much of it interlaced with the state's eight huge Carey Act projects envisioned by promoters. There, Volney T. Hoggatt, whose tenure as register of the state land board ended in early 1915, spearheaded his "Great Divide" dry farming experiment, which he located between the Little Snake and Great Northern Carey Act Projects. His method of development was to heavily advertise the vacant homestead lands as ideally suited for dryland farming. Hoggatt went all out to promote his dryland colony and by 1920 had enticed more than 1,000 dryland farmers to file on state and federal land in the region. But the influx was short-lived. Drought, a postwar price collapse of agricultural products, and ill-conceived use of the land given the region's sloping topography caused nearly every settler to abandon the land. Unfortunately, the debacle became Hoggatt's legacy.[9]

In March 1919 the three-person State Board of Land Commissioners of Colorado came to include—of all people—Will R. Murphy as its engineer member. The engineer turned newspaperman and one of the original founders of the Las Animas–Bent Carey project had run unsuccessfully as the Republican candidate for secretary of state in 1918. Newly elected Governor Oliver H. Shoup favored him with the state land board appointment, a position he would hold until his death in 1943.[10]

By early July 1919, Murphy and his fellow commissioners had confirmed that the Mutual Carey Ditch and Reservoir Company had sold water contracts to parcels of land that exceeded the mandated 14,000 acres necessary for construction to begin on the Muddy Creek irrigation system. Murphy optimistically reported that 240 prospective settlers had made payments to the trustee that amounted to more than $400,000. Accordingly, the state land board declared that monies in the hands of the trustee be made available to the company. The state land board approved a $250,000 construction contract between the company and local contractor William A. Colt, who had earlier purchased a water contract from O'Brien. To supervise construction of the Muddy Creek system, the company hired engineer Chalmers C. Schrontz of Denver. In August crews broke ground on the homogeneous earth dam using slips, fresno scrapers, and other horse-drawn equipment as well as hand tools to build the enormous dam in layers of dirt scooped from the reservoir site. The dam rose 50 feet in height and stretched 4,700 feet along its crest. Muddy Creek's lack of water likely prevented workers from moistening the layers to the extent that occurred at Two Buttes. The embankment consisted of almost 740,000 cubic yards of earth and 16,975 cubic yards of rock riprap that was 15 inches thick and placed by hand on both faces of the dam. The outlet consisted of three 4.5-foot-diameter riveted steel pipes laid in concrete and controlled by a battery of three cast-iron gates operated by hydraulic cylinder lifts. The spillway was at the west end of the dam and measured 300 feet wide. Outlet pipes, gates and gate controls, concrete work, tile drains, and a control house completed the 27,000-acre-foot–capacity structure. Workers completed the dam in November 1920 at a cost of $260,767.[11]

The state land board also approved additional monies for work on the Muddy Creek Outlet Canal to the segregated lands. Construction began in August 1920 with boring of its 1,473-foot tunnel through a sandstone spur 5 miles north of the dam. The project was challenging. The tunnel was 8 feet high by 11 feet wide, with a grade of 3 feet per 1,000 feet. The contractors

FIGURE 7.2. The Gerald G. O'Brien Carey Act farm located in the southwestern corner of the segregation (T.25S R.52W Sec. 20), ca. 1925. Author's collection.

excavated 9,000 cubic yards of solid rock and used 47,000 feet of lumber to complete the bore at a cost of $42,635, about $30 per foot. Meanwhile, workers continued excavating the main canal for 5 miles beyond the tunnel above Muddy Creek Valley. At a cost of nearly $83,000, the canal's excavation removed more than 216,000 cubic yards of earth and rock, 17,000 cubic yards of loose rock, and 9,000 cubic yards of solid rock.[12]

By the end of summer 1921, the Mutual Carey Ditch and Reservoir Company had spent more than $400,000. But it had yet to complete the Muddy Creek component. Plans called for the canal to continue from beyond the tunnel outlet and cross a company-constructed embankment for 4,000 feet to the southern half of the segregated lands. The canal was then to bifurcate around the Heinan Hills, where its branches were to join again to form the main canal that was to serve the northernmost segregated lands. But the company had run out of money and was waiting for reluctant colonists to continue making payments on their water contracts. In the meantime, about twenty families had settled on their Carey Act parcels and begun to dry farm while waiting for the canal's completion.[13]

Among those who had settled on the segregation were George E. O'Brien's just-married son, Gerald G., and his bride, Isabel Dodge O'Brien. Natives of Denver, the couple had purchased 160 acres adjacent to the

FIGURE 7.3. Isabel Dodge O'Brien, ca. 1917. Author's collection.

elder O'Brien's Desert Land Act tract. They arrived in the summer of 1921, as soon as roads and bridges reopened in southern Colorado following the catastrophic Arkansas River flood in June. To date, Isabel Dodge O'Brien's account is the only one that has surfaced about life on a Carey Act segregation in Colorado. Her perspective, while not the focus of the Carey Act's policy debacle, throws light on the frequently discarded aspects of daily routines that not only contribute to a farm's and ranch's productivity but also make rural life meaningful.[14]

From the beginning, Isabel O'Brien had doubts about buying into the Carey Act scheme and moving to the segregation in southern Bent County. However, she agreed to give it a try for ten years. Her misgivings were understandable. Thoroughly citified, she had a middle-class background and a few years of college under her belt when she and spouse (her term) spent several weeks preparing for the life-changing move.

FIGURE 7.4. On the farm. *Left to right*, Shirley, Richard, Barbara, and Geraldine O'Brien, ca. 1929. Author's collection.

Their drive to Las Animas and then south to their homestead was something of a shock, the lush farms along the Arkansas River suddenly giving way to barren desert-like stretches of what would be their new life. In the meantime, her father-in-law had built a temporary shack until *adobledoros* could complete the couple's new adobe home and outbuildings. The Hispanic-constructed dwelling differed from those of other settlers who generally built rock or frame homes, but it conformed to the custom of houses in Boggsville and elsewhere along the Purgatoire: "The house was pleasant with a living, dining room on the south. Several windows gave it a sunny exposure. The kitchen had a water supply, cabinets, woodstove, washing machine [was a] wringer and tub." The home had no electricity, and kerosene lamps sufficed as lighting. The outhouse was a ubiquitous feature. By fall, the home was ready for the couple to move in, and "it was a big day when a freight car of furniture, livestock, chickens, ducks and farm implements was unloaded [in Las Animas] onto wagons for the 18 mile trip."[15]

The arrival of belongings and farm equipment came at a critical moment—just in time for planting winter wheat—and set into motion the young couple's new life, which included both a division of labor and collaboration. They learned from day to day how to farm and ranch, and they coped with the frustrations of the incomplete Muddy Creek irrigation system. Gerald milked cows, ran cattle, tended sheep, dry farmed, and discovered the economic vagaries of agriculture. Isabel took on much of the domestic work, including doing housekeeping, laundry, and cooking, but an array of new jobs and responsibilities also became hers. Tending to chickens was an important task, gathering eggs as well as processing the birds. As she did with much of pioneering life, she took it with a sense of adventure: "For me that was something that I could do . . . We always kept a few of the heavier breed for baking and stewing." Occasionally, she also helped with the livestock, and at harvest time she cooked for the hired crew working the fields. But her greatest responsibility was nurturing the couple's children: "By the year 1925 we were blessed with four children, Barbara, Dick and the twins Shirley and Geraldine. I was a busy mother especially when the twins were babies. Jerry's mother helped at times by taking the older ones with her. His father was very fond of children so we were blessed with concerned parents." When the children were of school age, she took an active role in assisting at the one-room school, an adobe building with as many as a dozen students who learned reading, writing, and arithmetic. Their education was starkly different than her experience in the large Denver school system, and she compensated for some of the school's weaknesses with supplemental instruction.[16]

"As time went on," she wrote, "we learned to adjust to rural living. We loved the prairie and the open spaces." Spectacularly flaked arrowheads and other stone artifacts the family occasionally discovered on the property were reminders that other people had found utility and beauty in the land: "I'll never forget the blackness of the night, nor the howls of the coyotes . . . watching the longhorn cattle passing thru were [*sic*] fascinating." She and Gerald had rather quickly developed an emotional connection to the environment, an ethic, as the reader will see in chapter 9, that conservationists had long been seeking to enact into state and federal laws: "There was plenty of outdoor recreation for us. Good fishing holes on Rule Creek, fishing from a boat at the dam. Hunting for cottontails and quail. Trips to the cedars [south of the dam] for firewood. Later on family picknicks [*sic*] with the children."[17]

Writing in her simple prose, Isabel O'Brien concluded:

> All farmers must put up with crop failures for one reason or another. One summer day we stood at the window and watched a beautiful wheat field completely mowed down by hail. The dust storms were horrendous and to be expected every spring. . . . It was the winter of 1930 after Christmas we had a very heavy snow and freezing temperatures. There were ewes and lambs out on the prairie. Old Joe, the sheepherder, had stayed with them but most perished. . . . Two weeks later on Jan. 7, 1931, I stood at the dining room table looking out the window. There were swirls of smoke that seemed to increase. Our house was on fire. Luckily Jerry and his father and a hired man were close by. There was no way to save the house. The fire had started in the attic. Apparently, the flames had shot out thru a loose brick in the chimney. There was wind that morning. We had about 20 minutes to move out all that we could. Luckily we were all safe. It could have happened at another time and really been disastrous. Ironically we had lived there almost 10 years. We were some time getting over it. We moved to another farm northeast of Las Animas to begin a new life. We were 30 and 31 years old.[18]

The fire, though tragic, did not in itself end the O'Brien family's dream of an irrigated Carey Act farm. Isabel O'Brien did not elaborate about the decision to abandon the Carey Act segregation. But the family simply found it impossible to generate sufficient revenues from farming and ranching to cover their expenses. In any event, the ever-troubled Mutual Carey Ditch and Reservoir Company development constantly careened from one crisis to another as its directors desperately tried to salvage the project that seldom provided water from the ephemeral Muddy Creek. In addition, like farmers across the nation, the O'Briens found themselves suddenly caught up in US agriculture's massive economic collapse that began in the summer of 1920. As early as December 1921, settlers' debit balance owed to Mutual Carey totaled almost $100,000.[19]

Perhaps the settlers as well as George O'Brien, Frank Whiting, and other directors of the Mutual Carey Ditch and Irrigation Company believed US agriculture would quickly return to its golden period of economic growth, 1900–1914, and World War I's stimulus of high prices of farm goods. But the decline in postwar demand abruptly threw agriculture into an extraordinary economic depression that was worldwide in scope and lasted for decades. In Colorado, where farmland and commodity prices had reached unprecedented highs in 1918, the sudden plunge was dramatic. Farm bankruptcies skyrocketed as commodity prices dropped by as much as 45 percent. Across the state's eastern plains, several banks failed. An economic catastrophe had descended upon the state's agricultural sector, irrigated and dryland farms, ranches, and rural towns. The fundamental weaknesses that had crushed the industry lay in the reality that US agriculture had come to assume an inferior position in the nation's overall economy, which was becoming increasingly industrial and urban.[20]

Meanwhile, in early 1922 Mutual Carey's financial condition teetered on the brink of collapse as settlers, unable to irrigate from the incomplete waterworks, fell further in arrears on their water contracts. The company needed $40,000 to complete the heavy embankment for the Muddy Creek Canal and about $300,000 to build the Smith Canyon component but was in no condition to begin its financing stage. Distressed settlers and company officials pleaded for help from the state land board, which by then had canceled all but three of its Carey Act contracts: the Elk River Project in Moffat County, the Badito Project in Huerfano County, and the troubled Mutual Carey project. Only the constructed Two Buttes development offered an example of a completed Carey Act undertaking in Colorado.[21]

Yet remarkably, in what was a novel advocacy decision regarding Carey Act projects, the state land board determined in early 1922 that it would directly assist and lobby on behalf of the Mutual Carey Ditch and Reservoir Company. With Will Murphy heading the plan, the board and the company devised two strategic objectives: first, that the state of Colorado receive immediate patent from the General Land Office (GLO) to the segregation so that settlers could prove up their parcels from the state and, second, that the project reorganize as an irrigation district so it would be more likely to sell irrigation financing bonds.[22]

Those purposes of promoting irrigation schemes from flood-dependent streams, however, were at complete odds with definitive findings of the Colorado Irrigation District Finance Commission a year earlier. In 1919 the general assembly had authorized the commission amid an outcry over the decade-long inability to effectively market irrigation district bonds. The commission analyzed the history and condition of every irrigation district prior to drafting the framework of Colorado's 1921 Irrigation District Law for legislators. The new law added engineering analysis and a bond commission to the 1901 law and its various amendments and iterations. Of the state's sixty irrigation districts formed since 1901, just twenty-one were operational; all of those that were operational had earlier secured additional water supplies, adjusted acreages, and restructured serious bond debt obligations that allowed them to function efficiently. Most of those districts were in greater water supply regions, enterprises such as the Julesburg, Riverside, North Sterling (all original interests of South Platte Valley developer Daniel A. Camfield), and especially the San Luis Valley Irrigation District. The commission's report on the remaining thirty-nine inoperative, defunct, dissolved, illegally formed, in litigation, partly operating, or incomplete districts was scathing.[23]

The Colorado Irrigation District Finance Commission attributed much of the unmarketability of Colorado bonds to districts that had sought to reclaim benchlands by using floodwaters from intermittent streams on the plains of eastern Colorado. Near-mirror images of Carey Act projects, these enterprises were utter failures. The San Arroyo Irrigation District south of Fort Morgan, which sought to use an ephemeral tributary of the South Platte, exemplified the failings of dozens of private corporate projects as well. Formed in 1908 to irrigate 13,000 acres, the district sold more than $200,000 in bonds to Cincinnati investors and built its waterworks. But the system never received sufficient floods to supply its reservoir. Said the commission, "The district has served no good purpose, it would seem, except as a shining example (and so many were not needed) of the folly of attempting to impound flood waters of dry streams or arroyos flowing but perhaps once a year." An overestimation of thundering torrents and reservoir seepage losses doomed the project, and the financing bonds were in a constant state of default: "The future of this organization will be brightest which is shortest, and which secures a liquidation of debts and dissolution of the district."[24]

The state land board, however, more than just ignoring the findings of the Colorado Irrigation District Finance Commission, threw the full weight of its authority behind board engineer Will Murphy to advance the project's becoming an irrigation district. Murphy joined Mutual Carey president George O'Brien to develop five extensive reports that documented the project's status on the remote chance that the federal government might grant patent to the state before a permanent supply of water was available and that investors might find reason to finance its completion. The company hired the McMahon Audit firm of Denver to examine its books and produce a comprehensive financial statement. Engineer in charge Schrontz reported in detail the projected costs of completed construction; Addison McCune, state engineer whose personal involvement in the Carey Act was with the Routt County Development Company in Moffat County, expressed his "entire satisfaction" with the new arrangements.[25]

Murphy wrote that the project was viable and that the board's obligation was to help the efforts of "those men who have so persistently kept faith with the project . . . men who have colonized its lands and purchased its water could not be improved upon—good, practical, prosperous farmers of Colorado, Kansas, Oklahoma and Nebraska, for the most part. Two hundred and twenty-five of them have practically banded themselves together for the purpose of building this project mutually." Murphy emphasized that these people had "paid into trust something over $400,000, month by month. . . . Farmers these owners are, and their financial condition, first class and prosperous in the past, is now such that some kind of aid must be obtained if their original investment is to be saved . . . both the State and Federal Government must cooperate to assist these people."[26]

The state land board went immediately to Colorado's congressional delegation requesting that the United States Congress amend the Carey Act to permit the state a grant of patent before a permanent supply of water was available for the Muddy Creek settlers. Senator Lawrence C. Phipps, who sat on the powerful Committee on Irrigation and Reclamation of Arid Lands, and Representative William N. Vaile, who held a position on the House Committee on Public Lands, introduced identical bills to provide for the Carey Act segregation's patent. The men gained the support of Acting Interior Secretary E. C. Finney, who noted the case's "dissimilarity" with the ordinary Carey segregation: "In the opinion of the department this case presents such exceptional facts and equities as to warrant the issuance of patent to the State without awaiting the construction of the described additional facilities." The bills slowly moved through both houses of Congress. The only protest came from Rep. J. N. "Poley" Tincher of Kansas, who argued that the project would only worsen the dire water shortage conditions of his state's users of the Arkansas near Garden City and might impact ongoing litigation between Kansas and Colorado. In December 1922, ten months after the bill's introduction, Congress passed, and President Warren Harding signed, the law to benefit the Mutual Carey Ditch and Irrigation Company and its settlers. However, the GLO did not officially transfer the 24,404 segregated acres to the State of Colorado until October 26, 1923, ten months after enactment of the law.[27]

George O'Brien and William Colt had traveled to Washington, DC, to press their case, Upon their return, they successfully organized colonists to use their own labor to excavate mile after mile of the Muddy Creek Outlet Canal. By summer 1922 crews had dug the eastern branch of the ditch to within four miles of Boggsville. Crews also built the heavy embankment for the canal that crossed Heinan Draw. How the company paid workers is uncertain. Meanwhile, colonists also worked their small farms, utilizing dryland methods to grow modest yields of alfalfa, wheat, and kaffir corn. They also raised sheep, beef cattle, and dairy stock as a general optimism pervaded the segregation by the summer of 1923, when three solid weeks of storms drenched the region and filled Muddy Creek Reservoir. Moreover, a year and a half later, in April 1925, the general assembly approved an act that permitted the transfer of title to settlers who had fully paid their water contracts. Finally, it seemed that the company and stalwart colonists had a path to organizing as an irrigation district and utilizing the bond certification advantages of the 1921 irrigation district law.[28]

But any belief that financers might underwrite the remaining elements of the Mutual Carey project as an irrigation district was merely a false hope. The market for any irrigation bond remained generally unattractive to investors, despite Colorado's 1921 Irrigation District Law and those similarly revised by other western states. The Colorado General Assembly refused to enact the Colorado Irrigation District Finance Commission's two important recommendations to improve the marketability of bonds: one, that bonds be general assessment bonds that obligated districts to pay bonded indebtedness rather than the existing special assessment bonds and two, to note that the State of Oregon provided that its general revenues pay the first five years of a district's bond interest payments. Thus, by 1927, just four new irrigation districts in Colorado had formed under the new law. Three small districts reported having sold their financing bonds, but the 57,000-acre Trinchera Irrigation District in the San Luis Valley could not dispose of $750,000 in refunding bonds meant to consolidate prior debt and to expand and rehabilitate its system. State Engineer Michael Creed Hinderlider, who two decades earlier had helped engineer the Two Buttes Carey Act Project, summarized the bleak conditions for financing large-scale districts: "Due to the general depression effecting [*sic*] practically all agricultural pursuits it is virtually impossible to interest capital in the financing of new irrigation development, and most difficult to finance the requirements of many old and established projects." He saw no material relief in sight "until that inexorable economic law of supply and demand has had time in which to perform certain painful adjustments."[29]

Until farm revenues increased significantly more than farm costs, Hinderlider argued, capital for new private reclamation projects was unlikely. But inadequate revenues also plagued federal reclamation projects, where many farmers who had settled on government projects abandoned their homes and those who stayed were far behind on their payments. A generation of national investment in reclamation, United States Secretary of the Interior Hubert Work remarked, was at serious risk. Although just 6.3 percent of all irrigated land in the nation was that of the United States Reclamation Service (soon to be Reclamation Bureau), the government projects were among the most prominent features of all development efforts to transform arid lands. Work assembled a so-called Fact Finders' Commission that made sixty-five recommendations. Each suggestion bore the imprint of Elwood Mead, who had moved on from his role as head of California's Land Settlement Board; and each was an extension of the moral reasoning and practical recommendations he had made in Denver at the Conference of Western Governors in 1914. To correct reclamation's pattern of failure, Mead's longtime mission included congressional action to temporarily lift the required twenty-year time period for settlers to pay down Reclamation Bureau loans. This was among the most effectual suggestions Congress later implemented, as was the congressional write-off of $27 million in debt. Notably, before the committee had completed its report, President Calvin Coolidge named Mead to be commissioner of reclamation. Over the next five years, Mead implemented many of the Fact Finders' Commission's recommendations. Reclamation Bureau operations stabilized somewhat until the Great Depression disrupted them and the New Deal began expanding them.[30]

Debt forgiveness on private reclamation projects, in contrast, was at the mercy of the marketplace, where defaults bury failures. By the summer of 1927, the Mutual Carey Ditch and Irrigation Company had begun its own hemorrhaging. Fewer than a dozen families remained on their farms. Water in Muddy Creek Reservoir had mostly evaporated, but farmers managed to irrigate about 3,500 acres of crops, primarily corn and alfalfa. By November, the landholders, nearly $50,000 in debt, finally voted to dissolve their Mutual Carey Company and create the Bent County, Colorado Irrigation District. A year later the Colorado Irrigation District Finance Commission approved the district but limited the acres subject to bond issue to 23,511.[31]

The newly formed irrigation district authorized $600,000 of negotiable coupon bonds to build the Smith Canyon waterworks and to pay its indebtedness. The district also began issuing warrants to its creditors, an exchange permitted by public-authority laws to tax-permitting districts. By 1931 the irrigation district had issued $21,406 in warrants (notes carrying 7 percent annual interest payable when district-levied assessments were deposited with the county treasurer). George O'Brien was unsuccessful in his efforts to sell any bonds, and creditors would not negotiate for them. Moreover, from 1928 until 1937, the district made no assessment on its acreage to honor its registered warrants and the significant interest in arrears that mounted on them each year.[32]

George O'Brien and the remaining settlers were overseeing an insolvent irrigation district that painfully mirrored the circumstances of millions of businesses and other entities as the Great Depression crushed the nation. By the end of 1931, district farmers were cultivating fewer than a dozen parcels of the old segregation, none of which they irrigated because the reservoir was continually dry year after year as the great drought of the century began. Nonetheless, in 1933 the district's board of directors applied for a $500,000 loan from the Reconstruction Finance Corporation to build the Smith Canyon waterworks and fund its outstanding debt obligations. The loan program dated to President Herbert Hoover's administration. Its use for private reclamation projects showed the advancing acceptance of federal aid to distressed projects, arguably an inevitability with origins in the 1923–1924 Fact Finders' Commission. Federal officials, however, transferred the loan application to the newly created Public Works Administration (PWA) as part of the National Industrial Recovery Act, a law meant to pour money into the nation's economy through huge public works projects. By July 1933, the PWA had received a lengthy list of project proposals from the Colorado Committee on Industrial Recovery, a group of prominent businesspeople. Federal engineers quickly surveyed the various proposals, including the project headed by George O'Brien. In a matter of days, officials in Washington had given preliminary approval to more than 100 projects in the state—the Bent County, Colorado Irrigation District among them.[33]

Public Works Administration officials required, however, that before the Bent County, Colorado Irrigation District could receive the loan, the State of Colorado would need to enact legislation specifically making irrigation district bonds, at the option of the landowners, blanket obligation bonds. When the general assembly had enacted the 1921 Irrigation District Law, it had not included the investigating District Financing Commission's suggestion to do so. George O'Brien enlisted the help of Colorado legislators Representative Guy Hudson and Senator Fred Temple, both residents of Bent County, to change Colorado's Irrigation District Law to permit such taxing authority debt instruments and to allow a district's board of directors to levy ad valorem taxes on all land in the district. The Colorado General Assembly approved the general obligation bond measure on January 16, 1934. Within weeks, the Bent County, Colorado Irrigation District received notice that the PWA had approved an outright grant of $153,000 and a loan of $401,700 secured by a 4 percent general obligation bond of the district.[34]

The federal aid offered to the Bent County, Colorado Irrigation District was also contingent on its creditors subordinating their debt to the benefit of the PWA, which insisted on first standing for repayment. George O'Brien and the other directors may have believed that creditors would subordinate, that some arrangement might be struck. But the irrigation project's long history of failure had so infuriated the creditors that they refused to take a secondary position to the PWA loan. In early December 1934, US secretary of the interior Harold Ickes rescinded the loan and grant allotment. The decision killed any hope of expanding the system as originally conceived. The *Bent County Democrat* reported that the "big item of cost was the digging of the canal from the dam in Smith Canyon to the Muddy. The Muddy reservoir alone cost nearly $300,000 and it is estimated that approximately $500,000 has been spent so far on this irrigation project, all of which will be a total loss unless the district is able to sponsor some major financial program similar to the one the government proposed."[35]

By late 1937, the Bent County, Colorado Irrigation District's financial problems had become monumental. Its debt had ballooned to $94,540. Moreover, in August 1936, unpaid holders of district warrants had obtained a court judgment of $2,917 against the district. In October 1937 the district board of directors finally levied a special assessment on the lands totaling $3.92 per acre to pay all outstanding debt. For landholders still vested in the enterprise, the tax was crushing. The assessment far exceeded the worth of the land per acre. While those who owned about 4,000 acres paid the assessment, landowners with patent to more than 14,000 acres did not. Moreover, the state still owned about 6,000 acres.[36]

Since at least 1934, many landholders within the Bent County, Colorado Irrigation District had been furious about the condition of the project. Doctor A. B. Campbell, a retired physician from Las Animas who had paid $4,800 for his 160 acres with water rights, was among the angriest. In a letter to the GLO he unleashed a flurry of charges: "The State of Colorado has entrusted its obligations to incompetent and unscrupulous individuals, who have neither reclaimed the lands nor disposed of them to settlers." Campbell asserted that no water had been put to the land; that the construction company had wrongly foreclosed on numerous settlers' mortgages; that members of the board of directors held more than the allowed 160 acres per person; that Will Murphy, the engineer member of the Colorado State Land Board and one of the project's original founders, retained a financial interest in the project; that there would never be enough water to irrigate one-fifth of the 24,404 acres; and that at least 5,000 acres of the segregation were beyond the reach of the company's main ditch. He pleaded with the GLO commissioner to investigate the situation.[37]

The Department of the Interior took no action on Campbell's protest except to assert that the matter was a state issue, settled when the United States Congress approved the segregation in 1925. Other federal agencies also took no action after several aggrieved settlers petitioned federal authorities. Thomas J. Wilson of Loveland, Colorado, had purchased eighty acres, fenced it, and built a cabin. In June 1938 he wrote to President Franklin D. Roosevelt: "They left us the idea it would cost us $35.00 an acre for 1½ acres feet of water[;] now they tell us if the thing was finished we might get 5,000 acres irrigated on said land. I have put in $33.00 an acre . . . I am sending you a copy of the last assessment. This is what hurts." The White House directed the letter to the Department of Justice, which directed it to the Department of the Interior. The GLO sent the letter and others from brokenhearted settlers to the Colorado State Land Board, which filed them away.[38]

Meanwhile, George O'Brien continued to head the district. He left no written record about the conflict with settlers, and he revealed no reasons for staying with the disaster. But throughout World War II, except in 1943, above-average precipitation fell in the Las Animas area. O'Brien and several others irrigated annually no more than 2,000 acres scattered across five sections of the Heinan Flats. As domestic wartime farm production increased, land prices in Bent County began to rise slightly, and the acreage within the irrigation district came under new ownership. The Bent County treasurer began selling more than 14,000 acres acquired earlier for delinquent taxes. Moreover, on April 26, 1945, the State Board of Land Commissioners of Colorado sold its remaining 6,600 acres of Carey Act land in 160-acre parcels. Forty-two individuals, most from Bent County, each paid fifty cents per acre, the fixed price of Carey Act land. Tellingly, only one purchaser was Hispanic. The transaction was the final one for the state land board regarding the Carey Act in Colorado. All of the state-sold acreage and tax-sale land remained within the irrigation district. The district's 4,000 remaining acres belonged to George O'Brien, who over the years had purchased or acquired them by quitclaim title. In 1945 O'Brien, then seventy-one, suffered a serious heart attack. Broken and finally acknowledging defeat, he sold his 4,000 acres to Texas cattle rancher Grover Swift for $1 per acre. It was an ignominious moment. With the sale went a one-sixth interest in the district's waterworks. George E. O'Brien's futile thirty-five-year effort to develop an irrigation colony south of Las Animas mercifully ended, as did the pattern of Carey Act failures in Colorado. The Denver grocer turned developer had finally met his limits and retired to Denver, where he died in 1962.[39]

8

DEFEATED AND BURIED IN DUST

Fred L. Harris, lawyer, bond broker, and developer, began his eighth year as general manager of the Two Buttes Irrigation and Reservoir Company in 1917. He was forty-nine, still energetic, and he apparently believed the reclamation scheme could be successful because he continued to pour his personal finances into the project. Any hope for windfall profits, however, had disappeared from his mind, and the reality of a struggling enterprise was a constant concern. He still resided in Chicago, with work for various bond houses anchoring him there. Several times a year he traveled to Two Buttes and boarded at the town's only hotel. On occasion his wife, Anna, accompanied him. The couple, who married in 1898, had no children. Two Buttes was crude by big-city standards, but the couple found the hamlet adequate except for its lack of a railroad, a lament of most townsfolk.[1]

About 100 people had settled in Two Buttes. The town company, like the irrigation company, was the design of Fred Harris, Amos Newton Parrish, and Welley C. Gould. The town's first building was the hotel, a two-story brick structure complete with a restaurant. It also housed the company's

https://doi.org/10.5876/9781646426492.c008

FIGURE 8.1.
Fred L. Harris, ca. 1930.
Courtesy, Colorado State Archives, Denver.
Photographer unknown

headquarters. Newly arrived townsfolk, white and temperance-minded people who saw business or employment opportunities, built houses and storefronts on lots that comprised the quarter section–sized town, which incorporated in 1911. The Bank of Baca County, with Parrish actively involved, opened in June 1912. There was a livery, blacksmith, flourmill, drugstore, grocery, churches, and a newspaper—the *Two Buttes* (CO) *Sentinel*. Goods came from Lamar and sometimes from Springfield, by either wagon or truck. By 1915 the town had sidewalks (the first in the county) and had installed a gas streetlight system, a small electrical plant, telephone service, and a potable water system.[2]

While the town of Two Buttes may have shown signs of development, Fred Harris came to have serious doubts about the viability of the irrigation enterprise that straddled Prowers and Baca Counties—one of the two Carey Act projects in Colorado that did not seem to fit the failure pattern. But his frustration had grown with each succeeding year. Fluctuating reservoir water levels, variable annual precipitation, and colonists struggling to produce a sizable crop virtually overtook all hope that the project could be a success. However, like George O'Brien, he could not find it within himself to give up on the project. In 1914, when the impoundment held 29,000 acre-feet of water, it had been timely rains exceeding 18 inches that boosted crop production. During 1915, more than 20 inches of precipitation fell, but inflow to the reservoir measured just over 11,000 acre-feet. Late-season hail damaged much of the 5,860 acres of row crops. The year 1916 was dry, and the number of farmers totaled fewer than seventy-five—about the same number as had originally settled in 1911. Harris quickly learned how vital timely rains were to farmers' success under the system. In seven years of operation, the reservoir had never filled, making the water estimate of 1.5 acre-feet per acre a gross miscalculation. Thus, any idea that the 42,000-acre-foot-capacity reservoir could ever entirely irrigate the original 22,000-acre segregation quickly faded.[3]

The inadequate water supply was evident to the State Board of Land Commissioners of Colorado soon after the company had completed its waterworks because it relinquished back to the General Land Office (GLO) several parcels (about 2,000 acres of the original 22,000 acres) of the segregation to which it could not deliver water. But in February 1916, John E. Field, engineer of the state land board, traveled to Washington, DC, to present the United States Department of the Interior with Colorado's request for patent to 13,303 acres—the extent of land settlers had been able to cultivate, if not entirely reclaim. Field presented two filings. The GLO approved the first, a patent list of 11,511.36 acres that President Woodrow Wilson authorized in March. Two years later, in September 1918, the state land board relinquished back to the GLO 6,627.58 acres of the previously adjusted 20,000 acres for return to the public domain. The interior department then approved the second patent filing. President Wilson signed that grant to the state of 1,791.10 acres on February 14, 1919. What had been the hope of reclaiming 22,000 acres in 1909 became the reality of attempting to reclaim 13,303.[4]

The state land board was quick to transfer land patents to the Carey Act settlers who had proved up their parcels. But at the time of the 1916 patent, the project's total debt was untenable. Since 1913 the company's bonds had been in default, albeit by agreement with the bondholders that allowed farmers to delay principal payments by seven years with the execution of supplemental contracts (discussed earlier). By the summer of 1917, the more than seventy individual and institutional bondholders—including Homer W. McCoy and Company, which also held the principal ownership stake in the irrigation company—again accepted the reality of an investment that produced no returns. With most settlers unable to make any payments, the bondholders agreed to restructure the entire debt of the Two Buttes Irrigation and Reservoir Company. Fred Harris, his pessimism notwithstanding, agreed to facilitate the refinancing at the urging of Homer W. McCoy, who with other bondholders was unwilling to simply allow the company to collapse into bankruptcy and perhaps lose their entire investments. The parties agreed to the proposition of $662,816 in new bonds in exchange for a new deed of trust on the company property and virtually all of the settlers' hypothecated water contracts. In addition, the company gave Chicago Title and Trust and Harrison B. Riley 6,320 of its stock shares, which had a value of $158,000, making it and Homer W. McCoy and Company the principal owners of the Two Buttes Irrigation and Reservoir Company. Amos Newton Parrish remained president of the company and Harris the general manager. In just seven years (1910–1917), debt exceeded the entire worth of the enterprise.[5]

The parties believed they had placed a wager on the weather, not on climate, and that from the heavens might come the necessary precipitation to enable the segregation to thrive. But the fact that climate is a measure of atmospheric behavior over long periods of time and weather is a measure of atmospheric behavior over a short period did not register with most residents in the Two Buttes region. To be sure, sudden weather events did directly affect lives. During the second week of August 1917, just six weeks after the project's refinancing, a tornado ripped through the town of Two Buttes—killing one person, injuring several, and destroying buildings.

Also betting on the weather and the elements were people living across Baca County's vast 1.63 million acres on perhaps 600 scattered farms and ranches that drove the economies of towns like Springfield, Stonington, and Two Buttes. The great harvest of 1914 had started the influx of settlers to dryland farms. Hardy Campbell's disciples and others spread across the land. In 1916, Baca County experienced its greatest population growth. And by 1919, dryland farmers were cultivating more than 74,000 acres, with sorghum the largest crop at 43,000 acres, wheat at 17,000 acres, and corn at 11,700 acres. Broomcorn, soon a principal crop, lagged at 2,500 acres. Despite the region's arid conditions, its insect destruction, its per-acre harvest yields that varied with conditions, and farmers' difficulty moving crops to market without a railroad, the scale of dryland crop production dramatically outpaced production on the Carey Act lands. The price of wheat had risen to $2.02 per bushel, and the plow seemed to be the only commonsense implement across dryland fields throughout eastern Colorado. In 1919, a fairly good precipitation year at Two Buttes when 17.5 inches of rain fell, farmers cultivated little more than 9,000 acres but probably placed water on just a portion of the acreage.[6]

By 1920, the postwar collapse of agricultural commodity prices and land values had just begun. And by 1921, when the price of wheat had dropped to seventy-six cents a bushel, dryland farmers across Colorado began abandoning their places in droves. At Two Buttes, colonists were experiencing their own distress and began leaving the colony. By 1922, fewer than twenty-five colonists were actively farming, and a portion of those had rented out their places. Fred Harris, given the desperate conditions, intentionally took no action to foreclose on the delinquent and abandoned parcels. After a trip to Montana, where for several months he restructured the debt of the Big Timber Carey Act Project north of the Yellowstone River between Bozeman and Billings, he returned to Two Buttes in distress. In 1923 his wife, Anna, died. Adding to his anguish, the Prowers and Baca County project was effectively bankrupt, with no revenues generated from the enterprise.[7]

The unprofitable enterprise shared its lack of profitability with many completed Carey Act projects across the arid West, patterns of failure governors and reclamation experts left to the creative forces of the marketplace. Except for the Carey Act's application in Idaho and to an extent in

Wyoming, project after project was a financial failure. Actual reclamation of arid land in 1917 was no more than 700,000 acres of 14 million acres made available by the federal government. Nearly three fourths of the area irrigated under the act was land in Idaho. Of more than 100 Carey Act projects studied by United States Department of Agriculture analyst Guy Ervin in 1919, only three or four developments had returned profits to those who had financed them. The failings of developers were nearly universal: rainfall miscalculations, poorly selected land that was difficult to irrigate, and underestimations of the cost of construction and operation. Moreover, poor yields combined with these issues discouraged new settlers and made it impossible for existing settlers to make regular payments. Companies were often reluctant to foreclose on settlers for fear that the action would cause adverse publicity.[8]

Ervin thought many of the completed or partially completed Carey Act projects would eventually become cultivated enterprises, as either irrigation districts or state-controlled entities, or that they would be taken over by the United States Reclamation Service. He suggested changes to the law that included more accurate preliminary investigations, no settlement until a project was fully tested, supplemental contracts extended to all settlers, a policy that would allow only sufficiently experienced farmers to settle, elimination of the speculator by penalizing companies that failed to put the land into cultivation within a reasonable time, assistance for new settlers to help them develop their farms, and some means for making the cost of construction a lien on the lands by eliminating hypothecated notes on deferred payments.[9]

In 1920 the GLO began to more closely examine the feasibility of Carey Act project submissions. Although for years the number of new segregation proposals had been declining from its high in 1911, GLO engineers—those in the office and in the field—had begun to make more extensive examinations that included detailed feasibility reports on proposed projects. The new administrative method, reported GLO commissioner Clay Tallman, effectively ended new chimerical projects. In Colorado, except for the Two Buttes and Muddy Creek Projects, developers had abandoned the 1894 law.[10]

At Two Buttes, Fred Harris was overseeing the Carey Act debacle. Others would likely have fled the project, as had most settlers there. Time and again he had reorganized failed irrigation projects, though perhaps less challenging than the situation at Two Buttes. In part, his commitment to salvaging the project resided in his long friendship with Homer McCoy, whom he had known for twenty years. In 1921 McCoy sold his bond firm to Hill, Joiner and Company, also of Chicago, which acquired $30,000 of the Two Buttes Irrigation and Reservoir Company's non-hypothecated bonds.[11]

Harris was anchored to southeastern Colorado for other reasons as well. He had started developing his own farmstead on the outskirts of Two Buttes, and he had personally invested several thousand dollars in the irrigation project since its inception. Some of that money went to developing fishing at the reservoir, including experimenting with native catfish habitat and encouraging the Colorado Department of Game and Fish to introduce various species into the reservoir for public recreation. He also built a game bird farm. In 1925 he wed Mary A. Lewis, more than two decades his junior, and he adopted her two daughters: Selma, nine, and Frances, six. His new wife had been an early purchaser of a Carey Act parcel at Two Buttes. The family resided for a time in Colorado Springs before moving to Two Buttes in 1925.[12]

But professionally, for Fred Harris the move to Two Buttes was especially challenging. By 1927, the total of outstanding Two Buttes Irrigation and Reservoir Company bonds was more than $611,000. Creditors had agreed to defer principal payments for seven years in 1913, and after the parties restructured the debt again in 1917, the settlers made no payments on the principal. The only interest settlers had paid on any of the bonds was that on Series A. The last payment on that series was on July 1, 1921, at which time settlers paid about four-fifths of the face of the matured coupon. That total was just a few thousand dollars. The project's financial condition had become untenable. Since at least early 1924, Harris, ever the calculating developer, had been planning to force an involuntary bankruptcy on the company, a procedure whereby creditors filed the legal action rather than the debtor filing for protection against creditors. Chicago Title and Trust and Harrison B. Riley continually showed no interest in simply releasing the settlers' hypothecated water contracts executed with the Two Buttes Irrigation and Reservoir Company. Thus, the irrigation company, which was under the financial control of the trust company, was powerless to file voluntarily under the bankruptcy law. In June 1927 Harris wrote a blunt letter to bondholders and laid out his drastic reorganization plan:

> With a practical experience of eighteen years the water supply of this irrigation system has proved inadequate for more than about 3,600 acres of land. About 1,600 acres of this water is paid out, and the bondholders have no lien against it. This leaves about 2,000 acres available for the payment of bonds. Owing to the high cost of operation for such a limited area, the farmers on 2,000 acres, which is all that is being farmed at the present time, have been unable to keep the system in good repair and workable condition, and the wood stave pipes used to convey the water from the reservoir to the land are now in a state of complete decay. These pipes must be replaced before the abandoned water rights can be sold . . . I believe, however, that there is sufficient salvage to warrant the foreclosing of the property and a reorganization of the system on a smaller acreage basis, and to that end I am willing to undertake foreclosure, the repair of the irrigation works, and the resale of the abandoned lands, and turn over the proceeds, less the cost of foreclosure and repairs, to the bondholders who will deposit their bonds with me for that purpose. I shall make no charge whatever for my services, and shall require no advance of money from the bondholders in any event. My interest in the matter is to secure for the bondholders all the return that is possible, and to recover for myself some part of my investment in land in this project.[13]

Three months later, in answer to a bondholder's inquiry for further explanation of foreclosing on the Two Buttes Irrigation and Reservoir Company, Harris wrote:

> The history of this enterprise is a repetition of a thousand enterprises with irrigation. In this case the cause of failure was the shortage of water supply. Unless the property is immediately improved for service, titles straightened out[,] and the farmers provided with water so far as it will go the irrigation system will have to be abandoned, and if abandoned for any length of time will revert to a desert and cannot be revived with any salvage to the bondholders. On the other hand, if I am able to complete the foreclosure I have begun, clear up titles, repair the system and resell the abandoned water rights, it looks like some return may be secured for the bondholders. Just how much this return will be I cannot forecast. It might be ten cents on the dollar for the series A bonds, and possibly more. I am proceeding as economically as possible, and have brought the proceedings in a local court to save the cost and expense of proceeding through

> the Trustee and in the Federal Court. . . . Manifestly there's not present value in the bonds, and so far as I am able to see there's no prospective value except the salvage I may secure . . . I have been trying to save this enterprise for many years, and doing the only thing left to do with it. The original promoters and stockholders of the irrigation company are all dead or have abandoned it, most of the original settlers have likewise disappeared, the bond house which sold the bonds has gone out of business, and the irrigation system has fallen into such a state of decay that it can no longer serve its purpose. The bondholders are widely distributed and either indifferent or unable to take the initiative. I have myself lost a great deal of money on the undertaking, which I have long ago written off, though I still possess a great deal of land here which may be worth something if I am able to reorganize the system. My interest is identical with that of the bondholders.[14]

Thirty-six bondholders with Series A bonds totaling $117,250 joined Harris in bringing the deficiency in payment suit against the Two Buttes Irrigation and Reservoir Company, Chicago Title and Trust and Harrison B. Riley, and more than fifty non-depositing bondholders. Harris, in seeking the judgment, brought the action in the District Court of Baca County and hired the Lamar law firm of Gordon and Gordon to represent him and the other plaintiffs. Harris had continued to seek the cooperation of Chicago Title and Trust and Harrison B. Riley, but the house that held the deed of trust on all of the project's property and the water contracts refused to answer Harris or the court. A frustrated Harris thus attached the firm as a defendant in the summons. Writing to Henry H. Pahlman of Hill, Joiner and Company, who personally owned the large bundle of bonds that once belonged to Homer McCoy, Harris was exasperated: "I don't want to live in this damned desert forever, and would like to clean up the deal and get away from this eternal wind, dust, heat, cold, drouth, thorns, rattlesnakes, and squirrel-tooth nesters."[15]

Fred Harris secured his decree from the District Court of Baca County on May 21, 1928. It discharged all of the debt. At the sheriff's sale on June 16, Harris bought the irrigation system and all the hypothecated water contracts for $4,733. The enterprise was a liability, not an asset, and no one else bid. Nearly all of the costs were attorney fees charged by Willard and

Arthur Gordon. Harris reconfigured the irrigation enterprise by paring it down to 3,100 acres. He foreclosed on water contracts that embraced thousands of acres, canceling most. He resold 1,439 acres at their unpaid balance, usually about $25 an acre. Those acres, when added to the 1,647 acres owned outright by settlers, comprised the water supply capacity of the restructured enterprise. In addition, he began foreclosing on abandoned lands, reselling parcels when he could. Some went for as little as $2 per acre, the minimal price required by the state land board. He meanwhile began repairing the dilapidated siphons. His receipts totaled $38,088; expenditures, $24,372. The net salvage on $117,250 in bonds was a measly $13,716 (10 percent), with $5,862 distributed to the bondholders in cash on December 1, 1929, and the 5 percent balance due when collected from those individuals who had purchased the newly sold water contracts. Fifty water right owners, including Harris, comprised the reorganized project. Harris continued to serve as trustee of the enterprise and did not relinquish his oversight until December 1936, when the water right owners formed the Two Buttes Mutual Water Association. After twenty-six brutal years, the enterprise was finally debt free and its water rights secure.[16]

The involuntary bankruptcy in 1928 had lifted one burden from the backs of the remaining colonists at Two Buttes, but the profound economic and environmental forces of the 1920s and 1930s were inescapable burdens. Around Two Buttes, across Baca County, and throughout rural Colorado, farmers may not have understood the intricacies of the Great Depression that descended on them, but they certainly struggled with its consequences. Well documented by many historians, popular writers, and filmmakers, the cataclysmic economic collapse and killer dust storms combined to devastate the region.[17] The circumstance of the common farmer in Baca County enduring such conditions was unimaginable by modern measure. For instance, the family might still live in a house built of rocks, a simple frame house, or a dugout home in the side of a dune. The family was without electricity and often had no running water. The outhouse was still in use. In December 1927 Fred Harris wrote to an old friend, Anna Loomis, who had been one of the early purchasers of a Two Buttes parcel and later sold it. He articulated what many of his neighbors could not:

> In 1920 Baca County, Colorado, had a population of about 8,750. Today we have about 4,000. Montana lost 150,000 people between 1920 and 1925. Idaho lost 75,000 to 100,000 people in the same period. This depopulation is evidently going on in all parts of the United States in the agricultural states except the Pacific Coast. As a nation we are losing about 600,000 farmers, net, every year. These folks are, of course, going to the cities. And this city-ward movement is not only nationwide, but it is worldwide. The world is being industrialized rapidly, and education, the spread of knowledge, the realisms of applied science, the conversion into common and universal use today of the luxuries of yesterday, are all driving and luring the farmer from the farm to the city. How long and to what ultimate degree this movement will continue no one may predict, but probably it will be arrested in any event when the vast surplus [of] foodstuffs and other agricultural products declines to the point where it will permit a profit to be made by the agriculturalist large enough to enable him to maintain the standards of living of the industrialist. . . . In Illinois land values have fallen from $300 per acre to $125 per acre; in Iowa from $250 per acre to $75 to $100 per acre; and here in Baca County from $8 to $10 per acre to almost zero, and no buyer at any price. That is for the dry land; and most of our unimproved irrigated land may be bought for much less than the original price of $35 per acre, with all payments of interest and maintenance duly made. Men no longer buy land for speculative purposes, and only for actual operation and use; consequently they analyze a prospective purchase quite as much as intelligent men usually do other things. Formerly such matters as taxes, depreciation, upkeep, standards and cost of living, labor, material, transportation and price of product were not seriously considered nor carefully estimated in buying land. It was bought on the conviction that the increase in the value of land was always equal [to] or greater than the increase in the nation's population. This theory has been disproved by the fact, now known and recognized, that while the general population of the nation has greatly increased, the rural population is practically the same for 1920 as for 1900, and that since 1920 the farm population has diminished about six million. Here in Baca County we are experiencing on a small scale what the nation as a whole is undergoing.[18]

Harris, while acknowledging that the day of the speculator had passed, correctly grasped the dire position of US agriculture relative to industrial capitalism. However, he would never bring himself to acknowledge that economic regulation, which advantaged industrial capitalism, had any place in American agriculture. His entrenched belief differed from that of Elwood Mead and others who came to embrace the idea of federal subsidization as a means of aiding the nation's reclamation efforts.

Fred Harris's world was infinitely smaller than Elwood Mead's. But in November 1928 Harris took his laissez-faire economic philosophy to the state legislature when voters in Prowers and Baca Counties elected him to the Colorado House of Representatives on the Republican ticket. Perhaps Harris's decision to enter politics was related to the recent Lamar National Bank robbery and murder of its president and state senator Amos Newton Parrish, a Two Buttes founder. Parrish's son Festus was also fatally shot. In any event, Harris served in the Republican-controlled Twenty-Seventh and the Democratic-controlled Twenty-Eighth Colorado General Assemblies from 1929 to 1932. During the January following his 1928 election, he headed to Denver for three months. The committees Harris served on were Agriculture and Irrigation, Public Lands, and Appropriations and Expenditures, as well as, fittingly, Temperance. He lent his support to an old-age pension law, though it had no meaningful funding mechanism. As detailed in chapter 9, he notably sponsored legislation that made 7,680 acres on East Carrizo Creek in southwestern Baca County and 3,500 acres at Two Buttes Reservoir state game refuges. The designations authorized the Colorado Department of Fish and Game to manage the preserves for the benefit of wildlife. An avid sportsman, Harris would certainly have been aware of the activities in neighboring south-central Kansas to protect Cheyenne Bottoms, a huge wetland and waterfowl habitat. From the outset of his time in Denver, he did not expect to do much in the legislature: "There will be an exceptionally strong raid on the treasury by the bureaucrats, the schoolmen, the highway and road builders, and resistance to all of these is or will be my line." Returning home to Two Buttes in May 1929, he wrote to a friend that he considered his function in the body to be to "retard or defeat legislation [rather] than to promote it."[19]

The two terms Fred Harris served in Colorado's legislature were unimpressive. His tenure coincided with the beginning of the greatest economic calamity in US history: the 1929 stock market crash and the beginning of the Great Depression. His belief that government not only impeded free enterprise but also stymied individual initiative reflected his political philosophy as well as his moral code. It was a philosophy held by many Democrats in Colorado. By the first half of 1932, the Depression had buckled the state's already weak economy. The desperate conditions had so overwhelmed private and public charities that many closed their doors, and bread lines were common. A factor in the economic catastrophe was the dramatic slump in the agriculture industry, already weakened by a decade of declining prices. And the sector was beginning to experience the historic drought that would last through the 1930s. As one historian of the Depression in Colorado put it, "Either fields lay barren, parched under the hot drying sun, or crops rotted because harvest costs exceeded market prices. Like mining, manufacturing, and trade, agriculture had become a sick industry."[20]

Colorado received no federal funds to assist the destitute until 1932, and it was not until President Franklin D. Roosevelt took office that aid began flowing into the state. Fred Harris, running for a third state house term, lost a close election to Guy Hudson, a Democrat from Wiley; the vote totals were 4,446 for Harris to 4,798 for Hudson. By March 1933, Colorado's economy had almost come to a standstill. In Two Buttes, Fred Harris settled into the family house at the north edge of town. Since 1931, drought and more frequent dust storms had stressed his newly planted orchard of apples, peaches, and apricots, as well as a game farm he had built. His interest in party politics continued, and he attended the 1936 National Republican Convention in Cleveland as an alternate delegate. However, he never again sought public office.[21]

Fred Harris was not completely indifferent to the misfortunes of his neighbors in Baca County, but their condition had truly turned bleak. During the summer of 1931, the county poor fund was empty. The first federal help for Baca County arrived in March 1932. The local Reconstruction Finance Corporation (RFC) administrator distributed the funds for highway work, and the following year a second RFC grant provided $177,606 for roadwork. In October 1933 the Federal Emergency Relief Administration (FERA) added another $363,000 for roads. The Civil Works Administration under FERA began the first of ten construction projects in 1933 that employed more than 200 men. Nearly all the federal aid went to residents in the southern end of the county, a circumstance that disturbed townspeople in Two Buttes who sought aid for their region. They presented their case to county and state officials, but their complaints did not send specific aid their way. Nevertheless, federal assistance employed at least 1,100 Baca County residents on various work projects during the year. The amount of outside money promised or spent that year was $650,000; in comparison, Baca County's total crop income in 1933 was $755,000. Moreover, during that year, about one-third of the county's residents depended on federal relief for their subsistence. Those who were completely distraught began leaving the county in waves. By 1940, about 40 percent of Baca County residents had moved away. At Two Buttes, the downsized irrigation enterprise and smaller farmsteads surrounding it were the only areas in the county not drastically impacted by depopulation. But even there, the more fortunate could not escape the cataclysm.[22]

The Dust Bowl was a remarkable and terrifying phenomenon. Writing to Anna Loomis in December 1929, Harris had foreseen ominous signs: "There has been quite a boom in dry land in the county this year, and it still continues. But the buyers are all wheat men from Kansas, and they prefer a full section to a half section, and a half section to a quarter. They will buy a quarter occasionally, but an isolated eighty is almost unsalable. They are paying from $10 to $12.50 per acre, but it must be smooth and without breaks or depressions, so that it may be farmed with a tractor and harvested with a combine." Harris explained pensively:

> There are 1,650,000 acres of land in Baca County, and every acre of it is for sale of course, so that the price will hardly advance much higher before the natural condition of drouth dissolves the illusions which now promote the interests . . . everybody here is easily seduced into wheat farming, because it is less work, and this always breaks them. . . . One man out west of us, who was a Kansas wheat farmer, bought 3,000 acres of land a year ago in March, put it into wheat, and sold his crop for $71,000. He paid $16,000 for the land the year before. However, he probably will never get another crop of wheat, and sensing this he sold the place for $20 per acre.[23]

Two months later, writing to another of the original colonists, L. S. McMillan of Chicago, Harris elaborated on the boom in mechanized wheat farming around Two Buttes, across western Kansas, in the panhandles of Oklahoma and Texas, and in northeastern New Mexico:

> These buyers believe that by farming large tracts with power, and summer fallowing every other year, they can make wheat farming pay, even if the yield is not more than 10 bushels per acre. In my opinion they will be disappointed, because they certainly do not reckon upon the deficient rainfall or the violence and persistence of the wind. I have seen everything tried here in the last twenty years, and have yet to see wheat raised successfully. There was a large acreage of wheat sown last fall, but we have had a dry cold winter, and the winds are already beginning.[24]

The amount of Baca County acreage put into dryland wheat between 1929 and 1931 was truly remarkable, given that the price of the golden grain continued its drop from year to year. During the first half of 1929, more than 236,000 acres—one-seventh of the total area of Baca County—changed hands. During the first four months of the year alone, 1,116 land transactions occurred. That year, wheat farmers including Tom F. Hopkins, the Kansan Harris referenced, planted just 80,000 acres. In 1930 they put in 131,000 acres. In 1931 the acreage jumped to 237,750. Trapped by mortgage debt and falling wheat prices, farmers overproduced on a mega-scale. The dryland farmers' dilemma extended beyond the southern plains. Suitcase farmers, as the absentee wheat producers on the Great Plains came to be known, surrounded the town of Two Buttes. Meanwhile, perhaps two dozen colonists were irrigating 2,500 to 4,000 acres with skimpy flows from Two Buttes Reservoir. Some of the enterprise's farmers had done quite well, but they, like nearby wheat farmers, were at a significant disadvantage without a railroad for shipping crops. The Atchison, Topeka and Santa Fe Railroad finally arrived in the county in 1926, but it followed the wheat from Kansas and ran about twelve miles south of Two Buttes through the new town of Walsh. Privately, Harris admitted that it "looks very much as if I were something of an irredeemable idiot ever to have tried to work the confounded thing out. I would repine if I could feel certain that the project would eventually be a success even on a microscopic scale." He took no salary for his work. He had previously written off his losses and had just put his last $15,000 in cash into the deal when the unprecedented drought began in 1931. Yet he had faith: "There's an altar dedicated to hope on every quarter section of our shining desert, and someday our prayers may be answered."[25]

Harris's story, like that of many residents who remained in Baca County and in Bent County as well throughout the Dust Bowl region, was to be a chronicle of adversity. Instead of the enterprise's water supply improving, there was persistent decline. Writing in March 1935 to a bond investor who hoped for the last 5 percent payment on the bond she had deposited with him in 1928, he advised her to write the payment off as a loss. Successive years of poor crops and a continuation of the drought, which showed no signs of ending, made conditions bleak:

> Besides these evils we are now tormented with the most appalling dust storms, not heretofore troublesome, which damage our fields, retard or destroy vegetation, fill our ditches with sediment and menace human health and comfort. Since 1930 the total annual rainfall has not much exceeded ten inches, and in 1934 it was barely eight inches. . . . It is now doubtful if the irrigation system can, assuming an end of the drouth and a readjusted mean of supply and duty of water, furnish an adequate quantity for more than 2,500 acres of land with results that would bear the cost of maintenance and operation.[26]

Fred Harris disdained virtually every aspect of federal involvement in local economies, and, unlike George O'Brien, he refused to seek any federal support for the Two Buttes development. Although not openly hostile to the many New Deal efforts to stabilize the region's economy, he was indifferent to them. Among the most consequential of the efforts were those of the Agricultural Adjustment Act starting in 1933, which meant to raise farm incomes through regulatory controls; the Farm Credit Administration, which established a new credit system for indebted farmers; the United States Resettlement Administration (RA) in 1934, which absorbed and modified the FERA rural rehabilitation program; and the Colorado Soil Conservation Act in 1937. As analyzed in depth by historian Douglas Sheflin, the federal programs were filtered through county extension agents who lived in the various communities, resulting in transformative changes as Baca County farmers and ranchers significantly reduced production.[27] In 1933 farmers planted 50 percent less wheat than they had the previous year, and by the end of August, farmers had signed $236,000 in wheat bonus contracts. The cattle-buying program proved popular among the county's ranchers. Through 1935, 1,000 ranchers sold more than

19,000 cattle for a total of more than $265,000. The program also included the purchase of hogs and sheep.[28]

Harris seemed to have been less indifferent toward federal aid extended under the RA, which first surveyed the county in 1936 and found that farmers had abandoned or denuded 40 percent of its 766,000 acres of cropland. The RA concluded that 738,000 acres had eroded and needed remediation, a task local farmers alone could not accomplish. The soil conservation efforts that ensued took several forms, the most significant of which were the Colorado Soil Conservation Act and efforts of the United States Bureau of Agricultural Economics (BAE). In 1938 Baca County farmers voted to form two conservation districts, the Southern Baca County Conservation District and the Western Baca County Conservation District. That same year, the BAE began purchasing more than 200,000 acres of distressed land in the southern part of the county. With funds from the Bankhead-Jones Farm Tenant Act, meant to assist tenant farmers, not owners, the agency nonetheless paid the $34,000 in back taxes, which made the cost per acre $3.23. The BAE reseeded about half of the land before it eventually transferred all of it to the United States Forest Service, which manages it today by regulating grazing leases on it as the Comanche National Grassland.[29]

In the northeastern corner of Baca County, farmers did not approve the creation of the Two Buttes Soil Conservation District until 1942. But the stresses on both the land and the souls who lived there were enormous. Fred Harris, though living a more comfortable life than many, bore his share of burdens as the Depression deepened into the late 1930s. In early 1936 he wrote to an inquiring attorney who was unaware that the company's bonds had no value: "The long drouth of eight consecutive seasons has almost completely destroyed the water supply even for the reduced acreage, and the dust storms which continue to prevail threaten to obliterate the canals and ditches. I have been sitting upon the wreck for many years in the hope that there might be in time some further recovery, but that hope has entirely faded." He seriously believed "the enterprise was carefully, honestly and intelligently planned, but like any other adventure in our country certain elemental factors were not sufficiently explored, nor was the disastrous consequences [*sic*] of 'dry farming' on a large scale foreseen." In one of his last letters to Anna Loomis, in November 1937, he explained that "even weeds would not grow . . . we are constantly gazing upon a naked earth, hot, dusty and dead as the surface of the moon." In a rare moment, he agreed with the goal of having federal government agencies undertake erosion control.[30]

Between 1917 and 1944, the State Board of Land Commissioners of Colorado issued 128 patents to individual colonists who had fulfilled the proof requirement of cultivating specific units of their lands and paid in full their obligations to the company. But so many settlers had abandoned so much land that the irrigation enterprise was just one-tenth its original size. In April 1937, after Fred Harris transferred ownership of the irrigation enterprise from his control to the mutual water organization of about fifty farmers, he wrote to attorney Arthur Gordon: "I am not so sure that these people down here do not appreciate what I have done for them. You must remember that they have their troubles with the heartbreaking adversities of the climate, the drouth, dust, the insect pests, including the hookworm perhaps in some cases. The best of them, and there be some very good men among 'em, are rather strong for me, and have always been more or less appreciative. Some of these are not very articulate, and express themselves in other ways than words."[31]

Three years later, life's burdens had come to weigh particularly heavily on Fred L. Harris. His health had been fading in recent years. Chronic colitis and other ailments forced him to enter the hospital in Lamar on December 20, 1940. He died ten days later at age seventy-three. Funeral services were held at the Methodist Church in Two Buttes. Internment, however, was not in the buffalo grass, windswept prairie cemetery east of the town he had founded but in Lamar's Fairmount Cemetery. There, during the hot, dry summers, the well-irrigated Kentucky bluegrass stays beautifully lush.[32]

9

THE MAKING OF WILDLIFE AREAS

In the end, life's burdens and nature's limits overtook Fred L. Harris and George E. O'Brien. Their misfortunes at Two Buttes and Muddy Creek, the longest of Colorado's pattern of Carey Act failures, were heartbreaking. But despite the heartbreak—its delusion of windfall profits, rollercoaster financial markets, bankruptcy, drought and the Dust Bowl, grasshoppers, and conflict with settlers—each developer found some solace in appreciating the natural world in his reservoir creation. High sandstone cliffs, cedar-covered mesas, and long vistas background each location. As soon as the reservoirs held water, ducks, geese, and other seasonally migrating birds found the lakes; and both men—themselves outdoor enthusiasts—permitted waterfowl hunting. As mentioned in chapter 8, Harris personally developed fishing recreation at Two Buttes in 1914 when he encouraged the Colorado Department of Fish and Game to stock the reservoir. O'Brien welcomed small-game hunting at Muddy Creek in 1923. For residents of southeastern Colorado, the reservoirs, although artificial and private property, provided two of the very few areas in the region that

https://doi.org/10.5876/9781646426492.c009

wheat and broomcorn farmers had not plowed up and that anyone could enjoy. This local conservation interest in the reservoirs—the intermittent nature of their floodwater supply notwithstanding—reflected a decades-old outgrowth of state, national, and worldwide wildlife conservation efforts. It juxtaposed the profit-driven purpose of Carey Act development and celebrated a starkly different world—one that offered nothing of financial value but an abundance of aesthetic value, a wonder of each development's own making.

Harris and O'Brien saw beauty and splendor in that world, but neither developer could have fully appreciated the natural state of the streams before the dam building radically transformed the land. The impounded floodwaters of Two Butte and Muddy Creeks fundamentally altered native flora and fauna upstream and downstream, which for centuries had sustained Native Americans. Yet the reservoirs created their own constellation of life forms in the water column that extends from the surface to the bottom sediment and along the shoreline, as well as on land near the lakes. An inadvertent reciprocity, as one environmental historian describes the phenomenon, had sprung from the profit quest. As long as recreational use did not impact the business of irrigation, developers and water users accepted and indeed appreciated fish and other wildlife at both reservoir properties. Moreover, residents across southeastern Colorado came to regard Two Buttes and Muddy Creek Reservoirs as vital recreation spots and later led the effort that converted the purpose of the lakes from irrigation to wildlife conservation areas.[1]

Two Buttes Reservoir was chockablock full in 1914 when its first stocking of fish offered residents a different sense of the Carey Act project's place in their lives. Fred Harris and Alonzo L. Beavers, the General Land Office (GLO) register at Lamar and an enthusiastic fishing advocate, arranged with the Colorado Department of Fish and Game and the United States Bureau of Fisheries for the planting of 35,000 brook trout. The following year another 50,000 trout and at least 6,000 bass were stocked. Eastern Colorado had no state hatchery until 1932, so either agency would have delivered the trout from one of its mountain units. Trout, a cold-water-dependent fish, were a staple of reservoir stocking on the plains because so few warm-water species were available. The Colorado wildlife officials purchased warm-water species like bass from private hatcheries or made exchanges for trout with other states throughout the 1920s. Interestingly, Fred Harris favored stocking Two Buttes with channel catfish, a species native to the Arkansas River and its tributaries in southeastern Colorado. The survival success of all species at the reservoir and others on the plains was entirely dependent on the extent of irrigation drawdowns and, of course, drought. Nevertheless, fishing at Two Buttes, especially during the spring, was a favorite activity of residents of Prowers and Baca Counties. It not only brought people into direct contact with the world of nature but also became an expectation of life in the region.[2]

Meanwhile, in Bent County, George O'Brien welcomed waterfowl and quail hunting in 1923, although it is uncertain when the Colorado wildlife officials first began planting fish in Muddy Creek Reservoir. One report notes that the state agency granted the Bent County Game and Fish Club an allotment of ring perch and bass in 1929. The agency may have purchased the fish from a private hatchery in Nebraska or seined from other fisheries in the state. Therein lay the difficulty for O'Brien and the region's sportsmen's clubs that had formed to provide a stable fish culture across eastern Colorado. Transporting fish over long distances was costly, and mortality rates for all species of fish were high.[3]

Residents of eastern Colorado had long pressured the Colorado wildlife officials to build a warm-water-dedicated hatchery on the plains. The state's many cold-water hatcheries and the two federal hatcheries supplied mostly high-country streams and lakes, which also benefited the mountain tourist economy. In 1930, the Great Depression notwithstanding, officers of the Bent County Game and Fish Club—W. J. Frederick, H. R. Logsdon, Charles Hassinger, and Theodore Jorgenson—approached Roland G. Parvin, commissioner of the Colorado Game and Fish Commission, about the hatchery matter. The club subsequently raised enough money to fund construction of a hatchery. In the spring of 1931, work began on a twenty-two-acre site the department purchased along Adobe Creek northwest of Las Animas. The department hired Carl Peterson, who had run a small private hatchery near Wray, Colorado, to supervise the facility. Peterson expertly adapted the aquatic conditions of the new hatchery's rearing ponds for warm-water fish propagation. Beginning in the spring of 1933, deliveries from the hatchery stocked hundreds of thousands of hatchery-produced channel catfish, large-mouth black bass, white crappie, bluegills, and ring perch in eastern Colorado streams and lakes. Dust Bowl conditions and the extraordinary drought limited plantings until 1937, by which time the Colorado Department of Fish and Game had built a second warm-water-dedicated hatchery at Wray. Over the years, Two Buttes and Muddy Creek Reservoirs received large shipments of fish from these hatcheries.[4]

Fred Harris—a pensive observer and prolific letter writer—surprisingly left no record about fish stocking, the increasing numbers of waterfowl visiting the reservoir, or his role in creating the first state game refuges in southeastern Colorado at Two Buttes and at East Carrizo Creek in southwestern Baca County in 1931. Moreover, he left little record of his views on conservation in general, except his abhorrence of reckless dryland wheat farming. Even during the cataclysmic erosion years of the Dust Bowl, the Carey Act developer showed general indifference to the complex of federal- and state-incentivized soil and water conservation efforts under way by the late 1930s. His free-market leanings never changed, even though he had come to terms with the error of his enterprise. Perhaps too he rationalized that the dozen or so farmers irrigating parts of their parcels did so on just a fraction of the original design, making the scale of their practices less destructive. By 1937, just five of the original colonists remained, and not until 1941, a year after Harris's death, did farmers there form the 91,239-acre Two Buttes Soil Conservation District. In any event, the complex of soil conservation methods on the plains of southeastern Colorado represented its own turning point for the region's farmers and ranchers as they came to terms with the challenges of agriculture on the High Plains between the 1930s and the 1950s. And, as the reader will see in chapter 10, those transformative soil and water conservation actions directly interlaced with wildlife conservation development in the region.[5]

The staunchly anti-government Fred Harris was in his second and final term as state representative when he introduced his refuge bills in early 1931. State Senator Fred Temple, who represented southeastern Colorado, sponsored Harris's bills in the upper chamber. The measures sailed through the legislature without objection and became law on April 11. Like other state refuges, the Two Buttes State Game Refuge Law, in addition to providing 3,500 acres for the benefit of wildlife under the management of the Colorado Department of Fish and Game, was off limits to all hunting except by special license. However, it provided no protection to predators. In essence, the refuge designation was a perpetual use agreement that cost the company nothing and legitimized a dual purpose for the reservoir. The State of Colorado held the right to oversee wildlife, while ownership of the land below the reservoir and the land surrounding the water remained with the irrigation company, then nominally headed by Harris whose legal responsibility was to guard the company's 42,000-acre-foot water right adjudicated by the district court in 1922.[6]

The only notable accomplishment of Harris's tenure in the legislature was to increase the number of sanctuaries in Colorado to twenty-three, which comprised more than 3 million acres in 1932. The Baca County refuges were among just five the state had located on the plains. Only the Julesburg (Jumbo) Game Refuge, which the legislature created in 1927 in northeastern Colorado, was the site of another plains' irrigation reservoir. Since the creation of the first state game refuge in 1919, the legislative intent in establishing the sanctuaries was for the restoration and propagation of big game, such as elk, deer, bighorn sheep, and pronghorn antelope. Fish and birds were not unimportant in the argument for refuge designations, but habitat preservation for the large ungulates dominated the debate. Within the national forests, the amount of land set aside for big-game propagation was enormous, comprising hundreds of thousands of acres. Those great swaths, like designated private land, entailed no land acquisition cost to the state.[7]

Colorado's game refuges, however, had critics. Among the most outspoken critics in the legislature was Representative David C. Johnston, who chaired the house Game and Fish Committee during the Twenty-Eighth General Assembly and served on the committee in subsequent sessions. A Democrat from Golden, he was a constant critic of game and fish laws and of the Colorado Department of Fish and Game. Interestingly, he did not oppose the 1931 creation of the Two Buttes or Carrizo State Game Refuges. But in the following assembly in 1933, he erupted in protest during house debate on bills to create three new, very large forest refuges. During the daylong, heated debate, Johnston and other critics of setting aside additional land for wildlife argued that such creations were unconstitutional and that they unnecessarily removed vast hunting areas from the public.

He introduced proposals to reduce the size of several refuges and to abolish what he and others asserted were the most egregious examples of the law's abuse—the Two Buttes and Carrizo State Game Refuges.[8]

Supporting Johnston's effort was Representative Guy Hudson, who had defeated Fred Harris in the 1932 election. The Democrat from Wiley in Prowers County argued that the Two Buttes and Carrizo areas contained no game whatsoever. But leading the argument against Johnston's proposal was Representative Wayne Aspinall, a Democrat from Palisade, who was destined to become one of Colorado's most influential congressional representatives. Aspinall asserted that without such refuges, all of the state's wild game might well be exterminated within fifteen years. At the end of the contentious debate, the house agreed to redefine portions of two refuge boundaries: the State Game Reserve north of Rocky Mountain National Park and the Pike's Peak Refuge west of Colorado Springs. The attempt to outright abolish the Baca County refuges failed on final reading, and by mid-May 1933, in addition to preserving the refuges, the legislature had created five new game refuges. Coincidently, in June the Colorado Supreme Court ruled in *Maitland v. People* that the laws that set aside and created game preserves were constitutional. The court held that the state's wild animals, as well as the fish, belonged to all the people of the state and that the people had the right to set aside game refuges and preserve the wild game and fish therein.[9]

The state's high court ruling reflected a shared interest that was important to many people in Colorado and the nation, and indeed in other countries as well—wildlife restoration and conservation. Since before World War I, the conservation movement had grown rapidly, and its influence in public policy matters was significant. As mentioned in chapter 2, early conservation efforts in Colorado centered on the state's vertical landscape of mountains and western canyons as Americans were redefining their relationship with the natural world. Although those conservation efforts set aside unique natural areas, they ignored wildlife's place within them. Contrary to the thinking of those Progressive-era conservationists, Fred Harris never entirely disconnected the land from wildlife. However, following 1930, after more than two decades of contemplating the propriety of irrigating a near desert—albeit a place that once had abundant wildlife—he seems to have never joined any of the many sportsmen's clubs in the state that had come to advocate for wildlife restoration and conservation. The most prominent of these groups was the Izaak Walton League, a national organization of state chapters that exerted political influence on wildlife conservation issues.[10]

Such conservation groups were part of the evolving developments in American ecological resource management that directly linked wildlife with both nature and society. By 1930, as farmers were plowing up huge swaths of arid land on the Great Plains and farmers in the Midwest were draining wetlands to expand production, biologists had correlated declining numbers of waterfowl with those farm practices. In Wisconsin, as one example, wildlife advocates seeking bird habitat restoration at Horicon Marsh convinced the legislature in 1927 to create an 11,000-acre refuge. The action introduced alternative land-use options where agriculture proved inappropriate. The lack of federal funds for wildlife habitat at the onset of the Great Depression hampered advancing the land-wildlife link nationally. But the rethinking of wildlife management, as one historian has written, extended the scope of wildlife management, which came to reconcile political, ecological, and economic factors that thwarted species preservation. Such rethinking was the ideological framework of pioneering ecological spokesperson Aldo Leopold of the University of Wisconsin, whose land ethic inspired subsequent environmentalists.[11]

Colorado's wildlife conservation development, in contrast, related more to elk, deer, and pronghorn restoration than to birds. However, since the 1930s that development has always intertwined conservation-minded residents, game laws, and sufficient funding for wildlife purposes. During the territorial period, the legislature passed a number of game laws but provided no funding for enforcement. Upon attaining statehood, Colorado vested ownership of wildlife within its borders to its people in common, a doctrine that dated to the American Revolution and which the United States Supreme Court affirmed in the 1896 decision *Greer v. Connecticut*. State game laws and funding for wildlife purposes were slow to develop in Colorado: from a single fish commissioner in 1877, a fish and game commissioner in 1891, and a fish and game department in 1899, to a game and fish commission in 1937, which continues to administer wildlife conservation. In 1921, state-issued fishing and hunting licenses replaced all taxpayer funding of Colorado wildlife conservation; throughout the 1920s those fees funded game wardens and the establishment of state game refuges and special fish-rearing projects. However, by the 1930s, the ongoing Great Depression and years of drought had economically and environmentally devastated the state's fisheries and migrating waterfowl. Both big- and small-game populations declined significantly.[12]

Colorado's limited funding for wildlife restoration mirrored decades of national disinterest in funding devoted to wildlife. The great conservation and preservation efforts of the Progressive era that created millions of acres of national forest, monuments, and parks (at essentially no cost to the government) also created national wildlife refuges for migratory birds under the administration of the Bureau of Biological Survey of the United States Department of Agriculture. (The bureau also funded and operated federal fish hatcheries, including one in Leadville, Colorado.) By the early 1930s, the bureau's non-acquisition activities, such as international waterfowl protection, had become reliant on the sale of the one-dollar Duck Stamp, which the United States Congress established in 1934 and which waterfowl hunters were required to purchase. This national, hunter-based funding mechanism for wildlife conservation and restoration took another form in 1937 when the Seventy-Fifth Congress passed the Federal Aid to Wildlife Restoration Act, better known as the Pittman-Robertson Act or simply P-R. A monumental wildlife law, it redirected the tax on the sale of guns and ammunition (lowered from 10 percent to 8 percent) to exclusively funding state refuges—up to 75 percent of the cost per project, with 25 percent paid by the state.[13]

Colorado passed its Pittman-Robertson enabling legislation on May 10, 1939, although another decade would pass before the former Carey Act projects in southeastern Colorado received any P-R funds. Administration of the new law fell to the State Game and Fish Commission, which had replaced the Colorado Department of Fish and Game two years earlier. The commission was the product of legislative efforts to broaden the agency's representation and align its mission with the science of wildlife management, a methodology principally advanced nationally by Aldo Leopold. The State Game and Fish Commission hired Arthur H. Carhart, a confidant of Leopold's, to administer the P-R program. The state's first allotment of federal money was $22,438. As federal aid coordinator until 1943, Carhart oversaw P-R wildlife research and restoration projects, such as animal population surveys and propagation efforts that included the reintroduction of species to habitat-appropriate regions of the state where game had once proliferated. The restoration and protection of beavers in mountain streams was an important priority, principally for the animal's role in stabilizing the high wetlands across commodified watersheds. The federal tax continued to aid wildlife throughout World War II, since it was exempt from redirection to the war effort, unlike other excise taxes that went to the general fund. Following the war, federal aid to projects increased dramatically, and scientific management of wildlife resumed with vigor beginning in 1946.[14]

The Two Buttes Game Refuge received its first P-R funding in 1947, when the former Colorado Department of Fish and Game, now the Colorado Game and Fish Commission, designated approximately fifteen square miles around the reservoir as the first public shooting grounds for migratory birds in the state. State wildlife officials considered the area a unique wetland and its management an important part of efforts to preserve and develop such water environments, the loss of which was a nationwide occurrence. The Baca County Sportsmen's Club, headed by Roy Matthews, who was also a shareholder in the Two Buttes Mutual Water Association, must have welcomed the development. The company leased to the commission 3,062 acres for the nominal sum of seventy-five dollars per year. Wildlife officials dug shooting pits south of the reservoir.[15]

Meanwhile, a portion of the $34,310 granted for a comprehensive census of upland game and migratory waterfowl in Colorado went to study and implement a plan for managing the Short Grass Prairie Canada geese in the lower Arkansas Valley. At Two Buttes beginning in 1950, biologists and technicians began a decades-long assignment of trapping, banding, and conducting data recovery analysis on the unique birds. Under the direction of Colorado Game and Fish Commission researchers Jack R. Grieb and William H. Rutherford, ecological studies of the Arkansas Valley geese intensified as wetlands preservation, development, and acquisition became important objectives of state wildlife managers during the 1950s and 1960s. At Two Buttes Reservoir, technicians planted various grains for the resting geese, which regularly numbered in the tens of thousands. They also closely monitored hunting when 850 to 2,560 shooters per year visited the refuge between December and February. Because

FIGURE 9.1. The iconic landmark Two Buttes Mountain is private property adjacent to the wildlife area. Author photo, 2019.

of heavy silting near the dam, a residual pool of water usually remained for the birds' annual visit, even when irrigators drew down the reservoir and during the frequent droughts.[16]

In October 1957 the Colorado Game and Fish Commission, seeking to expand waterfowl hunting habitat, utilized P-R funding to purchase a 480-acre strip of land south of the refuge from Thomas and Cecile Kitzmiller for $16,800. For years, the couple had long supported developing the refuge. Meanwhile, five miles south of the town of Two Buttes, the migrating birds also began frequenting a 55-acre puddle created from a groundwater well located on property owned by the eccentric R. R. "Turk" Rutherford, a successful dryland wheat farmer who developed his property into a significant haven for goose hunters. The Colorado Wildlife Commission purchased his property in 1984 and established Turk's Pond State Wildlife Area—ironically, a true waterfowl refuge, closed to hunting from October 1 to March 1 every year.[17]

The drastic water fluctuations and siltation that had caused Two Buttes Reservoir to become shallow created critical water temperatures, which often made the lake a marginal fishery. Nevertheless, game and fish managers, responding to local pressure, planted 4.8 million fish between 1943 and 1960. Various limnological (inland water science) and fishery studies beginning in 1950 have detailed the challenges of creating a viable sport fishery. Early stocking included the warm-water species black bass, sunfish, crappie, yellow perch, drum, and channel catfish. Yet also at the insistence of local anglers, planting operations have included cold-water-dependent brown and rainbow trout. In 1955 managers began stocking walleye. Efforts to spawn the sport fish were rarely successful, with drawdowns and periods of drought reducing water to a level at which only non-game fish such as carp and the plains sucker survived. Nonetheless, stocking has continued at Two Buttes Reservoir as water levels permit.[18]

The science-based development of Two Buttes Reservoir as a fishery, like hundreds of others in Colorado and thousands across the nation, owed much of its financial support to the Federal Aid in Sport Fish Restoration Act, which the United States Congress passed into law in 1950. Better known as the Dingell-Johnson Act (D-J), its sponsors were Representative John D. Dingell, a Democrat from Michigan, and Senator Edwin C. "Big Ed" Johnson, a conservative Democrat from Colorado. The law revived a 1937 proposal to fund sport fishing by imposing a manufacture's excise tax on fishing equipment that had died during the Pittman-Robertson debate. Subsequent attempts to revive the tax met with bitter opposition from the American Sport Fishing Association, unlike the arms and ammunition industry, which supported P-R congressional passage of the Treasury Act in 1941—which dramatically increased taxes as war preparedness began and included a 10 percent excise tax on fishing tackle. However, the money collected went to the general fund and aided in financing World War II. The collection of excise taxes on fishing tackle into the general fund continued after the war, until finally, in the Eighty-First Congress, Dingell and Johnson managed to secure passage of the act directing the 10 percent tax to fund fish restoration and habitat development. President Harry Truman, who had opposed the bill for procedural reasons, came under increasing pressure from recreational anglers and

FIGURE 9.2. George E. Kimble, ca. 1953. Kimble was known to conservationists in southeastern Colorado as "Permanent Pool Kimble." *Courtesy*, Colorado State Archives, Denver. Photographer unknown.

reluctantly signed the Federal Aid in Sport Fish Restoration Act into law on August 9, 1950. Colorado assented to the act the following year.[19]

Dingell-Johnson funding proved transformative for Muddy Creek Reservoir. Years of neglect and damage by vandals between 1945 and 1950 had left the dam and its outlet gates inoperable. Heavy rains in 1950 filled the reservoir, and locals feared the dam might breach. In time, the water level receded through evaporation. The broken dam had long been concerning to those who lived downstream, including residents of Las Animas. Since at least 1948, Las Animas native William Setchfield Jr. had been advocating that the state should develop the reservoir as a fishery and wildlife preserve. Setchfield was president of the Las Animas Sportsmen's Club, which numbered more than 100 members. He enlisted the help of state representative George E. Kimble of nearby Swink to encourage Colorado wildlife officials to purchase and develop the area. The stocking of warm-water species from the Las Animas hatchery had made for good angling over the years when the impoundment held water. Moreover, scaled quail, which were native to the area, had become prized game birds of hunters who stalked the low canyon country. In 1953 officials of the Colorado Game and Fish Commission approached landowner Grover Swift about purchasing the reservoir.[20]

Subsequently, on January 3, 1955, the directors of the Bent County, Colorado Irrigation District, headed by Grover Swift, authorized selling the district's properties to the Colorado Game and Fish Commission. Following the sale, the irrigation district was dissolved. In early 1957 the Colorado Game and Fish Commission purchased all the water rights to the Muddy Creek and Smith Canyon systems and 2,438 acres of land that encompassed and surrounded the Muddy Creek Reservoir for $40,463. Seventy-five percent of the cost came from P-R and D-J funds, $22,760 and $7,587, respectively. The remaining money came from the commission's fund, made up mostly of hunting and fishing license fees. Swift shed his troublesome reservoir, and the Colorado Game and Fish Commission began its ambitious plan to develop a warm-water fishery and waterfowl hunting oasis.[21]

Thomas Kimball, director of the Colorado Game and Fish Commission, judged the cost to repair the pump house, dam, and spillway, as well as seeding dikes, building privies, and establishing fish populations, at roughly $213,000. To qualify for federal funds, soil investigations and hydrological studies were needed. The department hired engineer Orley O. Phillips of Denver, who concluded that rehabilitating the dam was an economical solution to developing the area as a substantial wildlife refuge. In June 1957 the department contracted with Spady Brothers of Las Animas to repair the dam, which the firm completed in July 1958 at a cost of $154,576. Kept abreast of the reconstruction was the Arkansas River Compact administrator Ivan C. Crawford. Since 1949, his agency had overseen all water development in the basin that impacted Colorado and Kansas. The Colorado Game and Fish Commission's most extensive

FIGURE 9.3. **Muddy Creek Reservoir's legacy and modern purpose comprises the Setchfield State Wildlife Area, about four square miles. Author photo, 2019.**

rehabilitation of the Muddy Creek Dam entailed reconfiguring and reconstructing the spillway—a central safety feature of the structure. The original spillway, built in 1919–1920, was an open channel carved in the natural earth formations, and it extended about 250 feet around the west end of the dam. Work in 1958 consisted of widening the spillway and pouring huge quantities of wire-reinforced concrete to strengthen its outlet to accommodate a flow of 28,000 cubic feet per second (cfs) if reservoir inflow ever required its use.[22]

The Colorado Game and Fish Commission rebuilt Muddy Creek Reservoir to a capacity of only 13,425 acre-feet, the original decreed water right. In April 1958, before the district court in Bent County, which was hearing appropriation matters, the department consented to the cancelation of 14,213 acre-feet of conditional right the Bent County, Colorado Irrigation District had never perfected. At the same time, the department consented to the cancelation of the conditional right of 340 cfs of the Muddy Creek Canal while maintaining ownership right to the original ditch decree of 60 cfs. However, it was not until 1963 that the Colorado Game and Fish Commission agreed to the cancelation of all its conditional rights to the Smith Canyon system—the canal's 300 cfs and the reservoir's 42 acre-feet.[23]

Muddy Creek Reservoir contained little water when fish biologists poisoned the lake to kill "rough fish species." By the summer of 1962, enough water had flowed into the reservoir that workers began planting white bass, crappie, walleye, and perch. Below the dam, several small impoundments provided seep-water storage and waterfowl nesting habitat. Also, Colorado Game and Fish Commission workers routinely planted eighty acres in sorghum to attract ducks and geese. Renamed Lake Setchfield in 1961, the wildlife area opened to the public in 1962. The only reference to its Carey Act past comes from locals who by habit continue to call it the Carey Dam.[24]

Never forgotten, however, was the month of June 1965 when the heavens unleashed a horrendous torrent across southeastern Colorado that killed twenty-two people and caused more than $560 million in damage. At a crossing on Two Butte Creek, floodwaters pulled Wilma Piper, age eight, and her brother James, age three, from the family station wagon. Their horrified father survived. Six months later, officials found the children's bodies at Two Buttes Reservoir, whose dam had miraculously withstood the flooding. Meanwhile, at Muddy Creek, floodwaters swept Mamie Blundell from her husband, James's, arms as they sought safety in their home upstream from Muddy Creek Reservoir. Searchers found her body eighteen days later in what remained of the reservoir. Floodwaters had lapped over the spillway, and then the dam breached and sent tens of thousands of acre-feet of water thundering toward the Purgatoire. Engineers later estimated that the Muddy Creek channel flood flow was between 45,000 and 80,000 cfs.[25]

To assess the damage done to Muddy Creek Dam, estimate repairs, and offer a recommendation about rebuilding the dam, the Colorado Game and Fish Commission hired Denver engineer Harold E. Eyrich. In his report filed in November 1965, he estimated the cost of repairing the structure at $458,000 to $611,000, depending on spillway capacity and repairs to the county road below the dam. He concluded that the region's drought cycles and evaporation rates made it unlikely that the impoundment could gather or impound water for extended periods. Moreover, the Colorado Game and Fish Commission's water rights at Lake Setchfield were junior to those of other users downstream on the Arkansas. There, irrigators might call for water stored in John Martin Reservoir, a 340,000-acre-feet flood impoundment built by the United States Army Corps of Engineers between 1939 and 1948 and overseen by the Arkansas River Compact. Should the John Martin Reservoir not be releasing water, its water users could demand all the water from Rule Creek even in the wettest years. Eyrich recommended that the Colorado Game and Fish Commission abandon the project as an impoundment.[26]

Upon Eyrich's recommendation, officials of the Colorado Game and Fish Commission chose not to reconstruct the dam for a second time, but they did not rule out rebuilding it. Moreover, they decided to continue ownership of the 2,246 acres that constituted the wildlife area and to manage it as conservation and hunting acreage. Since the 1950s, the population of native scaled quail, bobwhites, and Gambel's had increased dramatically in southeastern Colorado. Wildlife officials began managing brooding and covey sites on nearby Lake Setchfield as part of their greater management and transplanting efforts to boost quail populations statewide.[27]

Meanwhile, the Colorado Department of Fish and Game still possessed the decree of 13,425 acre-feet of water right in addition to 60 cfs of decreed canal water right. It was not about to forfeit those rights. Downstream at John Martin Reservoir the department and local wildlife groups had been seeking to establish a permanent pool for wildlife and recreation, something not provided for in the law that authorized construction of the immense flood control structure.

In southeastern Colorado the fusion of wildlife advocacy and policymaking after World War II had no greater example than state representative George E. Kimble of Swink. His advocacy for establishing a wildlife refuge at Muddy Creek Reservoir was merely one of many examples of his conservation efforts. Kimble was a retired railroad agent and organizer of the Holbrook Lake Sportsmen's Club, which affiliated with the Southern Colorado Council of Sportsmen and by 1960 numbered 5,000 members. From 1951 to 1956 he served in the Colorado House of Representatives; among his committee assignments were Game and Fish and, later, Natural Resources. He was a Republican and a strong supporter of the Colorado

Department of Fish and Game's scientific activities, which ramped up as increasing P-R and D-J monies began flowing into the state. Returning GIs interested in hunting and fishing across the United States fueled much of the increase in the two federal funds. Kimble also supported the department's purchases and leases of land for establishing wildlife areas, an activity the Colorado Department of Fish and Game resumed after the war. By the late 1950s, the state had purchased nearly thirty wildlife areas.[28]

Every life has compelling moments. For George Kimble, those of greatest historical consequence occurred when he was in his seventies. After he left the Colorado legislature he took an active role with the Colorado Wildlife Federation (CWF), an affiliate of the National Wildlife Federation (NWF). The NWF had spawned from the Game and Fish Protective Association but was also an outgrowth of wildlife restoration and habitat protection lobbying efforts that led to enactment of the Pittman-Robertson Act. However, it was not until 1952 that interested parties founded the CWF, which, along with the Izaak Walton League, were the most prominent non-government wildlife conservation organizations in Colorado. By 1960 Kimble was a CWF state vice president working out of Swink, where the Holly Sugar Corporation operated one of its sugar beet factories. Up and down the Arkansas Valley as elsewhere in the arid West, the industry consumed huge amounts of water. Irrigating farmers contracting with sugar beet factories and other farmers regularly drained their storage reservoirs. Kimble, his cohorts in wildlife conservation across southeastern Colorado, and state game and fish authorities accepted this reality as an everyday matter, given that sugar beets underpinned much of the region's economy. Fish losses and wildlife impacts were frequent in all of the valley's dried-up reservoirs, including Two Buttes and Muddy Creek. Incomprehensible to hunters, anglers, and recreationalists, however, were the troubling conditions that had existed since the late 1940s at John Martin Reservoir. When the United States Congress authorized its construction as the Caddoa Project in 1936, the lawmakers made no provision for a permanent pool for wildlife. After the United States Army Corps of Engineers built the reservoir, the United States Fish and Wildlife Service, the United States Department of the Interior, and the Colorado Department of Fish and Game began advocating for a permanent pool.

In 1952, under extraordinary pressure from outdoor enthusiasts, downstream irrigators permitted the use of their water for an experimental 1,000-acre-foot pool. That small amount of water proved completely inadequate for establishing wildlife habitat, and conservationist continued to advocate for a significantly larger pool for waterfowl.[29]

The permanent pool controversy at John Martin Reservoir pitted irrigators against outdoor enthusiasts. George Kimble had begun to vigorously support a permanent pool in the late 1950s. Writing in *Outdoor America*, the national publication of the Izaak Walton League, Kimble argued for a place in the reservoir to store Colorado water and indicated that the United States Army Corps of Engineers had reported its willingness "to assign 10 to 20 thousand acre-feet of space" for wildlife purposes. Moreover, he noted that the Colorado Water Conservation Board (created in 1937 to establish and maintain overall policies and procedures regarding water resources) had taken up the idea. Congressional action that might have addressed the issue stalled. But in 1964, in his capacity with the CWF, Kimble revved up the organization's publicity machine to back the permanent pool. Staunchly against the notion were the downstream Amity Mutual Ditch Company and US senator Gordon Allott, formerly Amity's attorney. Thousands of people, however, backed the permanent pool. In the aftermath of the 1965 flood, the United States Congress finally passed Public Law 89–298 providing for a permanent pool of 10,000 acre-feet, as long as Kansas and Colorado agree to its establishment under the Arkansas River Compact.[30]

Meanwhile, the Colorado Game and Fish Commission, seeking to preserve its water-use right, petitioned the Bent County District Court in 1968 to transfer 5,000 acre-feet of the department's storage right from Muddy Creek Reservoir to John Martin Reservoir as a permanent pool for fish, wildlife, and recreational purposes. Colorado water law provided for storage right transfers with a new water right, provided the shift did not injure the interested parties. But all of the lower Arkansas's senior water right users objected to the transfer: Colorado Fuel and Iron Company, Amity Canal, Bessemer Irrigation, Fort Bent Canal, Fort Lyon Canal, High Line Canal, Oxford Canal, and the Holbrook Mutual Irrigation Company. Also among the objectors was the Arkansas Valley Ditch Association (AVDA), which claimed the action would harm its rights. Judge

Lawrence Thulemeyer ruled that the transfer would injure the parties. But the objectors agreed to the change of location, provided that the state repair an existing measuring gauge in Muddy Creek and place another downstream in Rule Creek. They also insisted that the state engineer monitor both devices to measure water passing into John Martin Reservoir. With such remedies, the court granted the Colorado Department of Fish and Game the new storage right to in-priority use.[31]

In 1976, Kansas and Colorado finally agreed to the 10,000-acre-feet permanent pool, which they allowed to reach a maximum not to exceed 15,000 acre-feet. The following year the Colorado Department of Fish and Game petitioned for the transfer of its 8,425 acre-feet of storage rights still attached to Muddy Creek Reservoir to the John Martin Reservoir permanent pool. In addition, it sought a change of use right from irrigation and domestic purposes to recreational use. Principal objectors on this occasion were the Southeastern Colorado Water Conservancy District (which represented the interests of the major water users with senior rights) and AVDA. These objectors forcefully challenged the proposed transfer in Water Court, an authority created by Colorado law in 1969. Among a flurry of objections, the parties argued that the storage rights had been abandoned or otherwise promised by the Colorado Department of Fish and Game to be "set aside and forgotten" during the 1968 trial. On the one hand, they asserted that the reservoir had never regularly accumulated sufficient amounts of water to justify any further transfer. On the other hand, they stated that the Colorado Department of Fish and Game had not repaired and installed the measuring gauges on Muddy and Rule Creeks required by the 1968 water right decree. Judge John C. Statler found no evidence of abandonment or verbal promises and accepted the Colorado Division of Wildlife's (DOW) explanation of state budgetary issues regarding the inaction of water measurement devices. He nonetheless concluded that the objecting parties would incur injury without the specific remedies of placing, maintaining, and monitoring the gauges to not only measure flow but calculate transmission loss of water between the Muddy Creek and John Martin Reservoirs. With such conditions, the court issued the new decree on November 2, 1979. The action did not preclude the possibility of ever transferring the right back to Muddy Creek Reservoir if the state reconstructed it. Year in and year out, Muddy Creek has seldom made a significant contribution to the pool; both its inherently sparse flow and its junior water right are limiting factors. Today, Colorado Parks and Wildlife, as conditions permit, purchases water from several users to supply the pool.[32]

Meanwhile, at Two Buttes Reservoir, damage estimates from the 1965 flood amounted to only $57,667. The Two Buttes Mutual Water Association applied for an emergency grant offered by the United States Bureau of Reclamation. In August, the users received that amount and made necessary repairs to the dam's upstream face. For Roy Matthews, Joe Rose, Eldon Campbell, Forest Homsher, and the reservoir's other owners, the flood event could have been very costly. The flood hastened their longheld desire to sell the irrigation project to the Colorado Department of Fish and Game. In addition, as the reader will see in chapter 10, the revolutionary development of groundwater irrigation in the area factored into their desire to sell, as did the environmental movement's successful twin legislative accomplishments—the Wilderness Act and the Land and Water Conservation Fund (LWCF). For the Colorado Wildlife Commission, whose operational mission was to expand the protection of wildlife and offer opportunities to outdoor enthusiasts, the acquisition of the old Carey Act project neatly fit its policy objectives. Longtime state representative Forrest G. Burns of Lamar, known as the Prairie Fox, succeeded in obtaining legislative authorization for the Colorado Department of Fish and Game to purchase the reservoir and surrounding acreage. On August 13, 1970, twenty members of the Two Buttes Mutual Water Association sold the storied Carey Act debacle to the State of Colorado for $1,050,000. The commission paid for the purchase with a $525,000 federal grant from the LWCF and the remainder from the Game Cash Fund (moneys derived from hunting and fishing licenses). With the sale to the commission went the system's canal flow rights, its 40,918 acre-feet of storage rights, and 3,297 acres of land. All irrigation from William D. Purse's reservoir ceased. Water never again flowed down the main canal, through the tunnels, massive siphons, and numerous head gates to fields near the town of Two Buttes. Here too the Carey Act, like the story of the Muddy Creek Project, faded from people's memories.[33]

10

SUSTAINING A CONSERVATION PURPOSE

Nature's rhythms—the cycle of seasons, the turn of day to night—are the forces that pace the lives of wild animals that live in the aftermath of the Carey Act's pattern of failure in Colorado at Two Buttes Reservoir: the bluebird and the wild turkey, the mule deer and the pronghorn, the coyote and the cottontail rabbit, the bullhead catfish and the walleye. Some are residents, some are transients. The Canada geese, snow geese, and Ross's geese, for example, are birds of those consistent rhythms, ones with a remarkably tight social order. They begin their annual arrival at Two Buttes State Wildlife Area every November. The birds fly there and to other reservoir locations in the region from staging areas in Alberta and Saskatchewan. Canada geese (*Bránta canadensis*) start their great southern migration along the Central Flyway from nesting grounds even farther north in the Mackenzie River Basin. By late December at Two Buttes Reservoir, their population combined with snow geese (*Chén caeruléscens*) and Ross's geese (*Anser rossii*) frequently numbers more than 10,000. They usually leave the reservoir in the morning to feed in distant, privately

https://doi.org/10.5876/9781646426492.c010

FIGURE 10.1. Canada geese on Two Buttes Reservoir. Author photo, 2019.

owned stubble fields of wheat, sorghum, or corn and return in the evening. Wave upon wave of V-shaped formations and coordinated honking Canada geese and snow geese, as well as the smaller Ross's geese, are everyday sights in Baca County until late March when they begin their return migration north—and thus the cycle of life begins again.

The first person to systemically identify the birds of Baca County was amateur naturalist Edward R. Warren, who had an affiliation with the museum at Colorado College in Colorado Springs. During the months of April and May 1905, a year before Dr. William D. Purse began surveying his reservoir sites on Two Butte Creek and before the second influx of dryland farmers into the region, the well-respected wildlife expert and Colorado Springs resident found the avian world remarkably diverse and especially abundant throughout the undeveloped county. Although he saw no geese, it was spring, and nature's rhythms were pacing other birds northward. Visiting various ranches and farms and presumably public domain land, Warren saw thousands of birds, many at various waterholes and along the ephemeral and intermittent streams that constitute the county's few watercourses. He identified eighty-four species, most of which he trapped in cages or killed with a shotgun—the customary collecting methods of the time. He found ducks (teal, shovelers, coots), shorebirds (snowy egrets, sandpipers, killdeer, plovers), quail, prairie chickens, wild turkeys, various hawks, owls, woodpeckers, jays, and dozens of songbirds (sparrows, swallows, larks, buntings, warblers, and thrushes). Warren remarked that his list of birds would have been much longer had he devoted more observation time to them, but his priority was collecting mammals, his area of greater interest.[1]

The mammals Warren "collected" in southeastern Colorado were various rodents (rabbits, gophers, prairie dogs, and mice) and at least two predators: a bobcat and a coyote. He included observations about them in his *Mammals of Colorado*, published in 1910—coincidently, at the time settlers began purchasing Carey Act parcels at Two Buttes. Not unlike the anecdotal and observational methods of early military expeditions to the American West such as those of Zebulon Pike, Stephen H. Long, and James Abert, as well as later government railroad surveys, Warren's work reflected skillful detail and factual description. He paid little attention to biological theory, which had become the dominating school of thought dating to the Darwinian era and was the commonly held view of other mammologists of his day, including academics as well as scientists with the United States Biological Survey. Nonetheless, Warren's observations document an environment less disturbed than that in which Canada geese thrive in the twenty-first century.[2]

Edward Warren's 1910 magnum opus described the mammals he collected in southeastern Colorado by folding them into his larger picture of mammal distribution in the state—the first comprehensive volume based on exhaustive fieldwork that threw light on the condition and plight of higher-order animals across the mountains and plains. Long gone was any semblance of the Great Plains Serengeti, the homeland of Native Americans for centuries. He reported that the last bison in southern Colorado was killed near Springfield in 1889. Elk, once plentiful on the plains, were nonexistent except for a few in remote mountain valleys, as were pronghorns. Mule deer and whitetail deer, Warren reported, were declining in number, though temporary bans on hunting them, as well as elk and pronghorn, seemed to prevent their all-out annihilation. Very few grizzlies remained, and only in the mountains; Warren was delighted to report that in 1821 a silvertip killed Lewis Dawson, a member of the Jacob Fowler expedition, at the mouth of the Purgatoire. Some species, despite the development of agricultural lands, fared better. Wolves and coyotes, he noted, were present throughout the state. Bobcat and mountain lion populations were found in mountains and canyon areas and, surprisingly, in southeastern Colorado.[3]

The rhythms of the animal kingdom, however, were not particularly important to Warren. He eschewed biological theory in favor of data, as one biologist notes, yet he followed Charles Darwin's prejudiced belief that false facts were injurious to the progress of science because, unfortunately, they endure. Warren provided meticulous facts about animals over the course of his fieldwork, from 1900 to 1940. And he published ninety-six papers, leaving a legacy that proved essential to planning and protecting Colorado's wildlife. As Aldo Leopold wrote in 1949, "The art of land-doctoring is being practiced with vigor, but the science of land health is yet to be born. A science of land health needs a base datum of normality. . . . The most perfect norm is wilderness." Thus, Warren provided a carefully documented description of the normality of his time, a depiction of an environment where, if possible, preserving the remnants of Colorado's biotic diversity and restoring lands to greater health came to be significant aspects of modern environmental conservation—even on the plains of southeastern Colorado.[4]

Leopold's dream of an American wilderness and the efforts of other environmental advocates came to fruition in 1964, sixteen years after his sudden death, with enactment of the Wilderness Act. The law's authority to create wilderness areas within federal lands offered a promise of far-reaching wildlife habitat protection and restoration. Inextricably intertwined with the profoundly consequential Wilderness Act was the equally important Land and Water Conservation Fund Act (LWCF), the objective of which was to create a permanent federal fund without using direct taxpayer dollars to help pay for the preservation of outdoor recreation (and, later, cultural) resource sites across the nation—in wilderness areas, cities, small towns, and rural areas. As was its intent, the LWCF allowed government entities to purchase outdoor areas that not only included remote locations but also urban parks, baseball fields, swimming pools, and playgrounds—places where or close to where people lived. United States Representative Wayne Aspinall, who had championed early state wildlife areas while serving in the Colorado General Assembly during the 1930s, used his power as chair of the House Interior and Insular Affairs Committee to thwart a much more comprehensive wilderness proposal by effectively leveraging congressional consideration of the LWCF legislation, a bill he supported. Aspinall's version of the eventual wilderness law provided allowances for some industry uses within wilderness areas; together with the LWCF Act, it became law on September 3, 1964.[5]

Among the tens of thousands of LWCF project grants to federal, state, local, and tribal entities since the law's enactment was Colorado's use of the act to purchase Two Buttes State Wildlife Area in 1970. The dollar-to-dollar $525,000 matching grant covered half of the $1.05 million purchase from the shareholders of the Two Buttes Mutual Water Association. Wildlife officials paid the balance from their agency's Game Cash Fund, which came from hunting and fishing license fees—an accounting method they used repeatedly to expand wildlife conservation across eastern Colorado. At Bonnie Reservoir in far-eastern Colorado, LWCF dollars amounted to $1.8 million; at Queens Lakes north of Lamar, $60,000; at Lake Trinidad, $2.42 million. These purchases of wildlife areas and parks, in addition to other acquisitions such as acreage surrounding John Martin Reservoir, coincided with the development of ecological wildlife management in Colorado and the nation.[6]

After the Colorado Division of Wildlife (DOW) took ownership of the Two Buttes Reservoir property fee title in 1970, it continued to manage the

FIGURE 10.2. Two Buttes Valley with plains cottonwoods at the reservoir's rim where abundant wildlife lives year-round, though less so when the lake is dry. Author photo, 2019.

area—as it had for decades—as a fishing and hunting area. But seven years later, the division's director, Jack R. Grieb, took a direct interest in enhancing its 3,297 acres of wetlands. Grieb had been a principal researcher at Two Buttes during the early 1960s and was a renowned waterfowl expert. He designed new conservation outlays that included funding for personnel to improve the area's general wildlife habitat, including subsurface vegetative structures for fish and the planting of 100 plains cottonwood trees for raptors and other birds. Technicians installed two aerators to oxygenate the reservoir for aquatic life and to prevent freeze-over, thus allowing for the dispersal of waterfowl and preventing the spread of diseases among the wintering birds.[7]

By 1980 the Two Buttes State Wildlife Area, like all other Colorado state wildlife areas and natural area program land (acreage acquired since 1977 for endangered species and non-game habitat), was an ecological laboratory overseen by wildlife specialists. The Colorado Game and Fish

Commission had approved a comprehensive strategic management plan for all its properties that laid out general objectives: evaluating critical habitat for threatened or endangered wildlife species, managing species populations and enhancing habitat for game and non-game species, and increasing recreation activities. In southeastern Colorado at the Two Buttes, John Martin, and Queens State Wildlife Areas, as well as north in the Republican River Basin at Bonny Reservoir, the plan was to dramatically increase the population of a broad variety of species, both "consumptive and nonconsumptive." Area wildlife manager John Stevenson in Lamar and regional wildlife biologists Bill Olmstead and Chuck Loeffler coordinated with district wildlife managers as well as permanently stationed technicians at each wildlife area to implement the plan. The projects entailed about 40,000 acres.[8]

Jennifer Slater, the district wildlife manager in Springfield, along with wildlife technicians at Two Buttes, went to work on Olmstead and Loeffler's plans to sustain the property's purpose for wildlife conservation, both as a wetland and for the frequent times when the reservoir went dry. Thirty years of data gathering at the wildlife area gave the officials important insight into the ecosystem. Beyond structural additions such as fencing, signs, and markers, they continued to stock fish as conditions permitted. The technicians planted nearly 200 eastern red cedar trees across 15 acres of existing habitat and built an irrigation system to sustain their initial growth. Herbaceous plantings strategically placed for either wet or dry conditions were among the most important undertakings. These plantings included 100 acres of grasses, small grains, and legumes interspersed with native grasses and weeds to provide for feeding, nesting, and brooding plots. Usually off-limits to hunters, these dryland areas lure deer, waterfowl, pheasants, quail, wild turkeys, and a variety of songbirds. The work also included building wildlife ponds across the property to function as small playa-like desert waterholes. Over the period of 1979–1982, wildlife officials spent more than $65,000 at Two Buttes State Wildlife Area. The funding came from Pittman-Robertson (P-R) and Dingell-Johnson (D-J) programs and the Game Cash Fund generated from license revenues.[9]

Annual funding of the Two Buttes State Wildlife Area continues to enhance and maintain the unique ecosystem and wildlife's often precarious existence there. Between 1980 and 2010, wildlife managers purchased

FIGURE 10.3. Below the Two Buttes outlet tunnel is pristine wildlife habitat. Here, the area's wild turkey population thrives. Historically, the gallinaceous birds were resident in southeastern Colorado as late as 1906. Author photograph, 2019.

additional property that expanded the refuge to more than 8,500 acres. Beginning in the early 2000s, the vast flocks of Canada geese stopped wintering at the reservoir, a phenomenon also experienced at the John Martin and Queens State Wildlife Areas. Attributable to the birds now wintering at more developed national wildlife refuges across the Great Plains, frequent drought conditions in the Arkansas Valley, and newly developed irrigation reservoirs in Oklahoma and Texas, the enormous populations of *Bránta canadensis* have not returned. Also displacing the birds are new populations of the migrating snow goose and Ross's goose. These white geese are slightly smaller than their Canadian cousins. Sufficient reservoir water permitting, they often winter at Two Buttes, as do other migrating waterfowl that include varieties of ducks and similar swimming birds such as grebes, loons, and cormorants. By a Birding in Colorado estimate, more than 350 distinct species of birds visit or inhabit Baca County, and birders have spotted most of them at the Two Buttes State Wildlife Area.[10]

Decades of habitat enhancement and wildlife protection at Two Buttes State Wildlife Area provide a guarded space for nature's rhythms, but decades of soil conservation efforts across southeastern Colorado factored importantly in wildlife's twenty-first-century presence. Wild animals are creatures of movement, and their life cycles are often dependent on large and complex ecosystems. The combination of rapid settlement across the Great Plains, as Edward Warren noted, and the 1930s cataclysmic drought and soil erosion of the Dust Bowl era had decimated habitat and wildlife populations for hundreds of miles surrounding the Baca County refuge. The combination of New Deal policy responses to the erosion crisis and local landowner buy-in proved critical to wildlife's recovery or reintroduction in the region. Beginning with the United States Resettlement Administration (and later, the United States Bureau of Agricultural Economics [BAE]), the federal effort of land-use planning undertook ameliorating agricultural production on submarginal lands by purchasing unproductive land from willing farmers and resettling them to productive land elsewhere in the state. The BAE's efforts took at least 200,000 acres in Baca County out of production to rehabilitate the land and let it return to grass. Those acres, as mentioned previously, form the southern component of the Comanche National Grassland. Moreover, the United States Soil and Conservation Service, which offered federal money to farmers to participate in soil erosion control programs, assisted with contour farming to help prevent water runoff and blowing soil. It also helped with the planting of cover crops as part of crop rotation, constructed floodwater-capturing terraces,

FIGURE 10.4. **Here, the great siphons of the Two Buttes Canal once ran east then south for thousands of feet, pulling water uphill then down to the main canal that bifurcated to the segregated lands. Author photo, 2019.**

and planted tree and shrub shelterbelts. Finally, the all-important Colorado Soil Conservation Act (1937) has proved perhaps most important for private land conservation because it gave local control over that land to farmers and ranchers, which helped them balance economic production with soil conservation—with the assistance of federal money.[11]

The striving for healthy land and viable agriculture made for fewer farms and ranches but created larger operations, a reversal of federal land acts like the Homestead Act, Desert Land Act, and Carey Act. And it allowed state wildlife officials, with the support of conservation groups, to confidently begin to reintroduce numerous species of wild animals into the region, thus including wildlife as an important aspect of land health. Shelterbelts provided cover for pheasants and songbirds. Stabilized tracts of wheat, corn, and sorghum stubble enhanced goose feeding areas. The reintroduction of the pronghorn beginning in the 1940s represents one of the more remarkable success stories. As Edward Warren noted in his

1910 study of mammals in Colorado, the species had nearly disappeared from the state. The fastest land animal on the continent, the pronghorn, requires a vast range, and the soil conservation efforts in southeastern Colorado have enabled it to thrive in the early twenty-first century. Of the 80,000 pronghorns that live in Colorado, 10,000 roam across the region, and small herds frequent Two Buttes State Wildlife Area. Likewise, the reintroduction of mule deer, elk, and bighorn sheep has brought the rhythms of other wildlife back to the southeastern corner of Colorado.[12]

For all the effort to transform southeastern Colorado from its Dust Bowl cataclysm to a conservation-dependent region, Two Buttes Reservoir never escapes the agricultural purpose of its 100-year-old design. Wildlife managers had no reason to regularly operate the dam's outlet gates to release dangerously high water down Two Butte Creek because flooding seldom filled the reservoir. Nonetheless, the accumulation of silt, gravel, and weeds at the gates was a constant worry for managers. By one measure in 2009, siltation was as high as seventeen feet at the outlet gate. Interestingly, wildlife division engineers at that time, using new hydrology analysis, discovered that the original 1909 estimate of the reservoir's capacity level of 42,000 acre-feet was a drastic miscalculation. William D. Purse had based his survey of the high-water contour at an incorrect elevation. The later engineers concluded that the reservoir's correct capacity was just 23,322 acre-feet and that silting had reduced it by another 1,000 acre-feet.[13]

New hydrology analysis methods also helped engineers and wildlife managers address the hazard risk the dam had become if catastrophic flooding occurred. By the year 2010, silting had made the gates inoperative. Moreover, the spillway posed a risk of not diverting overflows. Few inhabited structures were located downstream to the creek's confluence with the Arkansas River at Holly, but a dam failure would likely have doomed the wildlife area to a catastrophe similar to the one at Lake Setchfield. To bring the old dam up to modern standards, John Clark, chief engineer for Colorado Parks and Wildlife (CPW; the successor agency to the DOW), and geotechnical experts used piezometers (subsurface sensing devices) to analyze the degree of needed repair. In 2012 workers built a cofferdam to separate reservoir water from the outlet gate so they could repair the inoperable mechanism. Later, workers removed tons of silt. They also removed phreatophytes, which included tamarisk, Russian

FIGURE 10.5. Heavy layers of silt in Two Buttes Reservoir, in addition to reducing the lake's capacity, show the challenge of safely maintaining the valve gates controlled from far atop the dam. Author photo, 2019.

olive trees, and invasive weeds that clogged the flow of water if released from the dam. Wildlife officials successfully tested the released water from the gates, which flowed into Two Butte Creek. The rusted and dilapidated 100-year-old equipment was working again. Yet all the repair work still left the dam vulnerable to failure. Repair of the dam is an ongoing task. In 2018 CPW engineers finished plans for a $3.4 million redesign to install new gates, a bulkhead, a gate control building, and a spillway tower and to improve the channel below the spillway. Travis Black, the CPW area wildlife manager, estimated that it would take years to secure funding and complete the project, one of 11 dams deemed critical of 101 owned by the agency.[14]

Just a small gathering of CPW officials and residents were on hand to celebrate the completed lesser repairs undertaken at Two Buttes Dam in 2012. The moment was unmistakably ironic. Generations of forgotten souls had come and gone since the failed hopes of irrigating 22,000 acres nearby under the Carey Act began with great fanfare in 1909. Yet astonishingly, in 2012 across Baca County, farmers irrigated about 80,000 acres—all of it watered from groundwater pumped to the surface and distributed through huge center-pivot sprinklers. The revolutionary technology transformed much of irrigated farming in Colorado and across the Great Plains beginning in the early 1960s. Across the land, green circles—some a mile in diameter—denote the Great Plains Aquifer of which the Ogallala is one formation. The pivots extend south and east of Two Buttes Reservoir, continue far into Kansas and Oklahoma, and irrigate wheat, alfalfa, sorghum, and occasionally corn. Dry farming and cattle ranching remain the main economic generators. Strikingly visible from Two Buttes State Wildlife Area is one of the large wind farms in southeastern Colorado that generate hundreds of megawatts of energy to Xcel Energy Corporation and the Tri-State Generation Transmission Association. Located on leased private property in Prowers and Bent Counties, hundreds of these towering turbines do supplement a few farm incomes. Capitalism remains omnipresent.

And yet, a conservation purpose continues to sustain Two Buttes State Wildlife Area, the lower Arkansas River, its southern tributaries, and the region overall. The complete commodification of nature—life's meaning for Joseph M. Carey and developers who believed it was possible to conquer the desert—yielded some to the forces of conservationism and that movement's later iteration of modern environmentalism. Colorado's 1937 enactment of the Water Conservancy District Act, which allowed for the creation of entities such as the Northern Water Conservancy District (mentioned in chapter 2), as well as the creation of the Colorado Water Conservation Board, which enabled centralized water planning at the state level, recognized wildlife and habitat as aspects of water policy but not especially significant ones. However, as Colorado's population on the Eastern Slope began to explode after World War II, and as the state's internecine water wars between the Western and Eastern Slopes raged, the formation of numerous water conservancy districts followed. In turn, environmentalists began to effectively challenge Colorado's water development; the halting of Echo Park Dam in Dinosaur National Monument was among the first of those successes.[15]

By the 1960s, environmental groups—such as the National Wildlife Federation, the Sierra Club, the Wilderness Society, and the Izaak

Walton League—had helped shift the national conservation about natural resources from commodification to environmental protection. Seminal federal legislation addressing rivers, such as the Wilderness Act and the Land and Water Conservation Fund, also included the Wild and Scenic Rivers Act (1968), the National Environmental Policy Act (1969), the Clean Water Act (1972), and the Endangered Species Act (1973). In addition to Colorado complying with federal laws, it passed its own responsive in-stream flow legislation in 1973. Those laws protected fishing streams, non-game and endangered species conservation, and aquatic nuisance species measures and created the Arkansas River Recreation Act, which authorized state regulation of recreational and commercial uses of the upper river. Subsequent state laws include fish and wildlife recovery rules related to water and irrigation development, and in 1997 Colorado initiated its wetlands protection program. The complex of federal and state laws underpins the management of all state wildlife areas, including those on Two Buttes and Muddy Creeks.[16]

Meanwhile, the remnants of the old Muddy Creek Carey Act Project, Setchfield State Wildlife Area—like all of southern Bent County—is a land without substantial groundwater. Since the dam's destruction during the 1965 flood and the decision not to reconstruct it, the property's principal wildlife and recreational asset—water—no longer exists. DOW officers managed its 2,438 acres as a shortgrass prairie that had a dry lake, juniper-covered hills, sandstone cliffs, and gullies. Hunters stalked the area for its scaled quail, mourning doves, and antelope. Bird watchers also sometimes utilized the area, particularly when the avian world migrated.

The two DOW transfers of 14,000 aggregate acre-feet of storage rights from Lake Setchfield to John Martin Reservoir, which dated to the 1970s, effectively redefined the location's wildlife management. Following DOW's long-standing policy objectives of providing recreationalists with greater opportunities and increasing wildlife and habitat in the region, wildlife officials entered a unique joint lease agreement with the Dean Land and Cattle Company that offered to satisfy DOW objectives, given the Setchfield property's lack of water. The agreement permitted public hunting access on a nearby 920-acre company tract located on the Purgatoire

FIGURE 10.6. Muddy Creek Dam's great gash of 1965 defines the Setchfield State Wildlife Area. Author photo, 2019.

six miles south of Las Animas in exchange for allowing the company to graze cattle at the Setchfield State Wildlife Area. The Purgatoire bottomland is an oasis, abundant with game such as whitetail and mule deer, waterfowl, wild turkeys, and other game and non-game species. Channel catfish and other aquatic life thrive there. The United States Army Corps of Engineer's construction of Lake Trinidad in the 1960s has effectively mitigated the Purgatoire's once catastrophic flooding and damage to the watershed.[17]

By 1980, local DOW officials had come to consider the Setchfield property surplus to the region's greater wildlife needs. In June they secured the Colorado Wildlife Commission's authorization to sell all 2,438 acres that comprised its designation. A month later officials advertised the property for "disposal" at the appraised value of $144,000. They received two bids but rejected them. In October the DOW again advertised the property, but this time for either sale or trade. Three parties offered bids. On November 20, 1980, the commission approved the proposal submitted by the

FIGURE 10.7. The Muddy Creek Canal at the Setchfield State Wildlife Area echoes a troubled past, though defined as a modern intermittent river and ephemeral stream ecosystem. Author photo, 2019.

Dean Land and Cattle Company. The offer included exchanging the company's deed to 940 acres of its Purgatoire property (appraised at $215,900) for the DOW deed to the Setchfield property, to which attached an easement deed for perpetual hunting and fishing rights. The DOW also agreed to pay property taxes in perpetuity on the acreage given its wildlife area designation, as it had done previously under the state's impact assistant grant, also known as the state wildlife payment in lieu of taxes. Edward Dean was a third-generation rancher in Bent County and a devoted advocate of wildlife conservation. His deed exchange proposal with the DOW included giving the agency a ten-year lease on 6,000 acres of his home ranch north of Las Animas.[18]

Complicating the exchange of deeds between the DOW and Edward Dean was the matter of the Setchfield property's 1957 funding with federal monies. P-R and D-J grants, monitored by the DOW's federal aid coordinator, required that project objectives meet wildlife and fish value formulas.

For unknown reasons, a federal review of the Setchfield property's circumstance and an analysis of the transaction occurred after the land swap. Following a January 1981 field evaluation by representatives of the United States Fish and Wildlife Service and DOW personnel, the officials compiled an extensive environmental assessment report, which concluded that the transaction—despite having originally ignored federal approval—was a seemingly equitable solution to regain lost fish and wildlife benefits that federal aid had principally underwritten. Distinctly different ecosystems, the Setchfield State Wildlife Area and the Purgatoire State Wildlife Area are something of an unknown world to most Coloradans and are among the few public land locations in southern Bent County.[19]

The fate of the four other Carey Act projects that Colorado developers sold to settlers is a story of mixed conservation outcomes. Neither the Ignacio Project in southwestern Colorado nor the Toltec Project in the San Luis Valley has experienced conservation-minded development. Both locations were relatively small development proposals. Lorenzo Sutton's Ignacio Project on the Los Piños is private property adjacent to the Southern Ute Reservation. Henry Clark's Toltec Project acreage that did not pass to protesting settlers remained with the General Land Office (GLO) until its successor agency, the Bureau of Land Management (BLM), began to oversee its use.

However, across the great swaths of land that were once the fiefdoms of the Little Snake River and Great Northern Projects in northwestern Colorado, the management of ecosystems is now a principal aspect of land use in a region that remains cattle and sheep country. A blend of state land board and BLM properties that ranchers lease, the vast area benefits from government planning resources that include detailed environmental impact studies and funding. More than seven square miles of the Little Snake River fantasy once planned by Addison McCune, Monroe Allison, and Warren Given now comprise the Little Snake River State Wildlife Area. An everyday sighting here are cottontails and sage grouse. The region is home to the largest pronghorn herds in the state, more than 20,000 animals by one count.[20]

Twenty miles east of the Little Snake delusion of a Carey Act project, the acreage of the Great Northern Irrigation and Power Project—the agency of David Moffat and later overseen by Lafayette Hughes—is a widely scattered grouping of wildlife lands. Elkhead Reservoir and State Park (built in 1974 by the DOW and a group of electrical power providers) traps water flowing from California Park, once the hope of Carey Act developers. More than $30 million in improvements to this impoundment make it a vital component in flow management for four endangered native fish species: the pikeminnow, humpback chub, boneytail, and razorback sucker. Four other state wildlife areas comprise the old segregation's remnants: 640-acre Cottonwood Creek, 104-acre Boston Flats, 1,122-acre Triangle, and 724-acre Maynard Gulch.[21]

The history of the Carey Act in Colorado is the story of a great disconnect between private developers' hopes for windfall profits and the reality of unstable financing, economic rollercoasters, and the physiographic challenges of reclaiming high sagebrush lands. Conversely, the history of wildlife conservation that has transformed those benchlands, most evidently in southeastern Colorado, is the story of connectivity among conservationists, state and federal agencies, and private landowners. The story continues. More than three dozen public entities and private organizations as well as some landowners are voluntarily partnering to conserve, enhance, and restore prairie and wetland ecosystems in the region. The CPW's role in southeastern Colorado conservation initiatives is the most prominent, given its management of about 60,000 acres that comprise 27 wildlife properties of the 355 statewide. Its force of scientific experts, access to various financial resources, and district officers embedded in local communities work closely with other environmental groups with varying funding capabilities that also play a role in preserving the region's ecosystems. One high-profile collaborative effort among the parties to conservation efforts, including local landowners, is the Playa Lakes Joint Venture, which identifies and funds intermittent water resource conservation projects. The Nature Conservancy also has a history of strategically purchasing ecologically important properties that contain intermittent rivers and streams, as in the Carrizo area watershed. The Colorado Natural Heritage Program at Colorado State University in Fort Collins is a major database of ecology-related resources used by those conservation groups. It links to other ecology-related clearinghouses across the United States and to the greater worldwide conservation of intermittent rivers and ephemeral streams, also known as IRES.[22]

Of the earth's vast network of river drainages, more than half are IRES, and climate change is accelerating their numbers. The value of IRES, both environmentally and economically, is the subject of extensive studies that document these terrestrial ecosystems across the United States, Europe, New Zealand, South Africa, Israel and Palestine, India, Brazil, and elsewhere. The IRES of Australia are the most studied. There, the complex relationship between organisms and dry streams demonstrates how these distinctive ecosystems are valuable national and world resources. Ecosystems are linkages of life. And those connections of life forms include microbes, algae, vascular and non-vascular plants, invertebrates, fish, amphibians, reptiles, and mammals. Australia celebrates its marsupials, and IRES studies have shown economic development's environmental impact on species that are dependent on temporal waterways for habitat and migration. Worldwide, legislative actions have set aside IRES conservation zones, though mostly in national parks. But even in Australia, IRES are not well understood, and there is no societal consensus about their environmental and economic value.[23]

Setchfield State Wildlife Area, despite its artificial playa-like lake, is an example of an ephemeral stream ecosystem worthy of protection. Its ecological value notwithstanding, the area is probably the least visited of Colorado's state wildlife areas. That is both a shame and a blessing. But a diversity of species not only passes through the broken-down dam site and adjacent hillsides and cliffs but also resides there. It is a recommended stop on the Colorado Birding Trail, but birders often overlook the old Muddy Creek Carey Act Project. It is nearly always a place of solitude. Pronghorn and mule deer occasionally pass through the area. Eastern bluebirds, pine siskin, canyon towhees, rock wrens, curved-billed thrashers, ladderback woodpeckers, juniper titmice, red-tailed hawks, and long-eared owls either visit or reside there. Also living here are reptiles of all sorts: horned lizards, rattlesnakes, bull snakes, and the western coachwhip that grows to seven feet in length. Amphibians include toads and salamanders. Many of the varied creatures, the pace of their lives synchronized with nature's rhythms, are undoubtedly part of the vast and remote ecosystem of the lower Purgatoire.

FIGURE 10.8. Ferruginous hawks hunt rodents from this single tree at the ruins of Isabel and Gerald G. O'Brien's home on the Muddy Creek Carey Act segregation. Author photo, 2007.

Thus, remarkable is the diversity of wildlife that exists and often thrives in the aftermath of the Carey Act's long pattern of failed projects in Colorado. Also remarkable is the shift in mind-set from the boundless-thinking developers to the passionate environmentalists that has led people to reconfigure their relationship with arid land and capitalism's exploitation of it. Fish, wildlife, and environmental considerations factor into virtually every aspect of water development and land-use planning in the state. The creation of the Colorado Water Conservation Board in 1937 has proven pivotal in wildlife conservation along waterways. Through its administration, federal and state directives as well as funds provide for fish and wildlife resources. Water conservancy districts are the most important vehicle through which wildlife conservation receives representation in water development. The Southeastern Colorado Water Conservancy District, which dates to 1958, and the development of the United States Bureau of Reclamation's Fryingpan-Arkansas trans-mountain project, which transmits 57,400 acre-feet of Western Slope water to

Arkansas Valley users, rationalize much of their function as maintaining and improving wildlife habitats. The Purgatoire River Water Conservancy District, formed in 1960 as a repayment entity to the federal government for constructing Lake Trinidad, also dedicates a percentage of its operation to fish and wildlife habitats. The Lower Arkansas River Conservancy District (2002), beyond its mission of protecting and enhancing economic water resources, seeks to preserve and develop wildlife habitat by facilitating Colorado's conservation easement program for preserving agricultural and habitat acreage for wet and dry lands. Since 2015, the Arkansas River Watershed Collaborative—a branch of the Arkansas Basin Roundtable (2005)—along with the Colorado Conservation Water Board, has served as a sounding forum for water development and wildlife issues in the region. The complexity, intricacy, and collaboration entailed in contemporary water policy in Colorado would have been unimaginable to the otherwise boundless minds of Carey Act developers.[24]

Rethinking the past, including the missteps of the Carey Act at Two Buttes and Muddy Creek, has served southeastern Colorado and its residents well. The public lands are treasures, as are the tens of thousands of acres of other preserves in the region. Among the most sacred of these preserves may be its cultural designations: Bent's Old Fort National Historic Site, which venerates the commodification of the bison and America's imperialistic ambitions during the Mexican-American War in the 1840s; the Sand Creek Massacre National Historic Site, which memorializes the deep meaning of one of the state's most horrific crimes when two regiments of Colorado volunteer soldiers murdered and mutilated 230 peaceful Cheyenne and Arapaho men, women, and children in 1864; the Boggsville Historic Site, which invites visitors to learn about Hispanic Colorado and the meaning of pioneering Hispanic women in the state's past; and Amache National Historic Site, which acknowledges the internment of more than 10,000 Japanese Americans during World War II and forces serious introspection about racism.

After a pilgrimage to the Sand Creek Massacre Site in 2012, I—a great-grandson of George E. O'Brien—traveled fifty miles southwest to the Setchfield State Wildlife Area, where I sat near the deep gash in the Muddy Creek Dam. Alone, I contemplated the irrigation project's sad fate and the greater meaning of the Carey Act's abject failure across the state. There had been malfeasance to be sure, and Carey Act developer O'Brien carried his share of blame for the disaster. But the story was essentially about the boundless nature of the mind. Just across the spillway, several head of cattle were grazing on the juniper-covered hillside where workers had quarried rock for the dam's riprap in 1919. Over my shoulder came a slow chat-like series of calls and trills. Bouncing atop the dilapidated valve house was a loggerhead shrike. Black-and-white feathers and a heavy-hooked bill identified this robin-size bird. It chirped a discordant song. Had I paused too near its nest hidden in a thorny bush below? Perhaps it was inevitable that the shrike should come to live there, as do swallows under a bridge. In any event, for reasons we may never know that are hidden in nature's rhythms, the uncommon bird had made its home in the forgotten shadow of a Carey Act project.

APPENDIX A

The Carey Act

Section 4 of the original act of August 18, 1894, entitled, "An act making appropriation for sundry civil expenses of the Government for the fiscal year ending June 30, 1895." (28 Stat., 372–422)

That to aid the public-land States in the reclamation of the desert lands therein and the settlement, cultivation, and sale thereof in small tracts to actual settlers, the Secretary of the Interior, with the approval of the President, be, and hereby is, authorized and empowered, upon proper application of the State to contract and agree, from time to time, which each of the States in which there may be situated desert lands as defined by the act entitled "An act to provide for the sale of desert land in certain States and Territories," approved March third, eighteen hundred and seventy-seven, and the act amendatory thereof approved March third, eighteen hundred and ninety-one, binding the United States to donate, grant and patent to the State free of cost for survey or price such desert lands, not exceeding one million acres in each State, as the State may cause to be irrigated, reclaimed, occupied, and not less than twenty acres of each

https://doi.org/10.5876/9781646426492.c011

one hundred and sixty-acre tract cultivated by actual settlers, within ten years next after the passage of this act, as thoroughly as is required of citizens who may enter under the said desert-land law.

Before the application of any State is allowed or any contract or agreement is executed or any segregation of any of the land from the public domain is ordered by the Secretary of the Interior, the State shall file a map of the said land proposed to be irrigated, which shall exhibit a plan showing the mode of the contemplated irrigation and which plan shall be sufficient to thoroughly irrigate and reclaim said land and prepare it to raise ordinary agricultural crops and shall also show the source of the water to be used for irrigation and reclamation, and the Secretary of the Interior may make necessary regulations for the reservation of the lands applied for by the States to date from the date of the filing of the map and plan of irrigation, but such reservation shall be of no force whatever if such map and plan of irrigation shall not be approved. That any State contracting under this section is hereby authorized to make all necessary contracts to cause the said lands to be reclaimed, and to induce their settlement and cultivation in accordance with and subject to the provisions of this section, but the State shall not be authorized to lease any of said lands or to use or dispose of the same in any way whatever, except to secure their reclamation, cultivation and settlement.

As fast as any State may furnish satisfactory proof, according to such rules and regulations as may be prescribed by the Secretary of the Interior, that any of said lands are irrigated, reclaimed, and occupied by actual settlers, patents shall be issued to the State or its assigns for said lands so reclaimed and settled: *Provided*, that States shall not sell or dispose of more than one hundred and sixty acres of said lands to any one person, and any surplus of money derived by any State from the sale of said lands in excess of the cost their reclamation, shall be held as a trust for and be applied to the reclamation of other desert lands in such State. That to enable the Secretary of the Interior to examine any of the lands that may be selected under the provisions of this section, there is hereby appropriated out of any moneys in the Treasury not otherwise appropriated one thousand dollars.

APPENDIX B

Carey Act Filings in Colorado

Compiled from the records of the Colorado State Land Board.

NO FILING NUMBER ASSIGNED. Pawnee Pass Irrigation Company, 1895, Dan Camfield and George West, 300,000 acres proposed near Fort Morgan and Sterling in Weld, Morgan, and Logan Counties, $2 million proposed revised to $300,000, no bonds issued, filing withdrawn 1902, no land reclaimed under act.

NO FILING NUMBER ASSIGNED. T. C. Henry Project, 1900, Theodore C. Henry, 10,000 acres proposed near La Junta in Otero County, no proposed dollar outlay, no bonds issued, filing withdrawn 1900, no land reclaimed.

NO. 1, A & E DITCH, 1902, C. E. Ayer and J. R. Ellis, 2,029.75 acres proposed near Craig in Moffat County, no proposed dollar outlay, no bonds issued, filing withdrawn 1903, no land reclaimed.

NO. 2, no filing.

NO. 3, LITTLE SNAKE RIVER (Colorado Realty and Securities Company, later Routt County Realty and Securities Company), 1903, M. L. Allison, Addison J. McCune, and W. C. Johnston, 38,000 acres proposed near Maybell in Moffat County, 38,000 acres segregated by United States Department of the Interior, cost of water $45 per acre, $200,000 proposed, revised to $750,000, $500,000 of bonds issued, $400,000 sold, filing withdrawn 1916, no land reclaimed.

https://doi.org/10.5876/9781646426492.c012

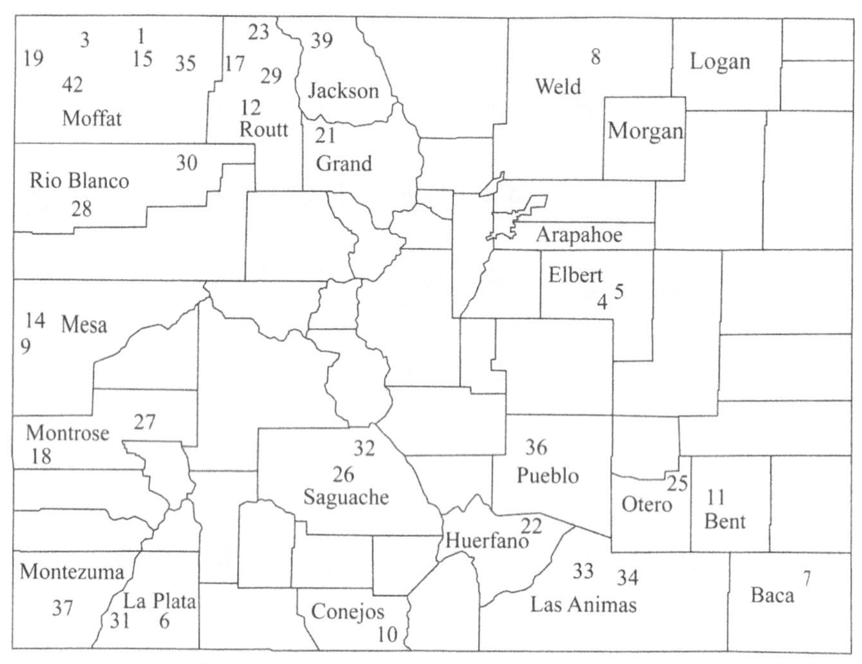

MAP 8. **Carey Act filings in Colorado**

NO. 4, DEER TRAIL LAND AND CATTLE COMPANY, 1904, 2,200 acres proposed near Deer Trail in Arapahoe County, no proposed dollar outlay, no bonds issued, canceled by state land board 1906, no land reclaimed.

NO. 5, NOONEN, John Noonen, 1904, 2,200 acres proposed near Deer Trail in Arapahoe County, no proposed dollar outlay, no bonds issued, canceled by state land board 1907, no land reclaimed.

NO. 6, IGNACIO (Colorado Land and Water Supply Company), 1906, Lorenzo M. Sutton, 16,000 acres proposed near Bayfield in La Plata County, 16,000 acres segregated by United States Department of the Interior, cost of water $45 per acre, $3 million proposed, no bonds issued, canceled by state land board 1915, no land reclaimed.

NO. 7, TWO BUTTES (Two Buttes Irrigation and Reservoir Company), 1908, William D. Purse, 22,000 acres proposed northeast of Springfield in Baca County, 22,000 acres segregated by United States Department of the Interior, cost of water $35 per acre, $700,000 proposed, $475,000 of bonds issued and sold, 13,303 acres patented by General Land Office to State of Colorado, 7,000 acres maximum annual irrigation.

NO. 8, PAWNEE IRRIGATED LANDS, 1908, Francis P. Lonergan, 12,000 acres proposed near Keota in Weld County, $500,000 proposed, no bonds issued, canceled by state land board 1913, no land reclaimed.

NO. 9, SOUTH PALISADE FRUIT, LAND AND WATER COMPANY, 1909, W. H. Nutting, Rod McDonald, and Lloyd Sigler, 13,908 proposed acres near Whitewater in Mesa County,

$250,000 proposed, no bonds issued, canceled by state land board 1910, no land reclaimed.

NO. 10, TOLTEC CANAL, 1909, Henry H. Clark, 14,852.74 acres proposed near Antonito in Conejos County, 12,000 acres segregated by United States Department of the Interior, cost of water $40 per acre, $3 million proposed, no bonds issued, canceled by state land board 1915, no land reclaimed.

NO. 11, VALLEY INVESTMENT COMPANY, 1909, Alpheus L. Pollard, 24,000 acres proposed south of Las Animas in Bent County, 24,000 acres segregated by the United States Department of the Interior, cost of water $35 per acre, $1 million proposed, $400,000 collected from colonists, $600,000 of bonds issued, none sold, 24,000 acres patented by General Land Office to State of Colorado, 4,000 acres maximum annual irrigation.

NO. 12, D. G. LEACH, 1909, D. G. Leach, 30,000 acres proposed south of Hayden in Routt County, $500,000 proposed, no bonds issued, canceled by state land board 1913, no land reclaimed.

NO. 13, same as List No. 26 (Stark-Hagadorn Irrigation).

NO. 14, TWIN RESERVOIR AND IRRIGATION COMPANY, 1909, unknown petitioner, 3,400 acres proposed near Collbran in Mesa County, no proposed dollar outlay, no bonds issued, withdrawn 1909, no land reclaimed.

NO. 15, GREAT NORTHERN IRRIGATION AND POWER, 1909, E. C. Phillips, H. R. Phillips, and James R. Kilpatrick, 240,000 acres near Craig in Moffat County, 142,732 acres segregated by United States Department of the Interior, cost of water $55 per acre, $1.5 million proposed, $150,000 expended, no bonds issued, canceled by state land board 1920, no land reclaimed.

NO. 16, same as List No. 30 (White River, Trappers Lake Reservoir).

NO. 17, C. O. NELSON, 1909, C. O. Nelson, 29,720 acres proposed north of Greystone in Moffat County, no proposed dollar outlay, no bonds issued, canceled by state land board 1909, no land reclaimed.

NO. 18, DUNCAN CHISHOLM (San Miguel Canal), 1909, Duncan Chisholm and Bulkeley Wells, 75,000 acres revised to 43,000 proposed west of Nucla in Montrose County, $4 million proposed, no bonds issued, withdrawn 1914, no land reclaimed.

NO. 19, BROWNS PARK WATER COMPANY, 1909, unknown petitioner, 25,000 acres proposed in Browns Park in Moffat County, no proposed dollar outlay, abandoned 1909.

NO. 20, same as List No. 26 (Stark-Hagadorn Irrigation).

NO. 21, W. H. FOSTER, 1910, W. H. Foster, 5,000 unidentified acres proposed in Grand County, no proposed dollar outlay, abandoned 1910.

NO. 22, BADITO (Huerfano Valley Irrigation Company), 1910, Alonzo P. Sickman, 36,779 acres proposed southeast of Pueblo in Huerfano and Pueblo Counties, $1.5 million proposed, no bonds issued, canceled by state land board 1924, no lands reclaimed.

NO. 23, NORTON MONTGOMERY, 1910, Norton Montgomery, 26,000 unidentified acres proposed in Moffat County, no proposed dollar outlay, abandoned 1912.

NO. 24, same as List No. 29 (Lafayette M. Hughes).

NO. 25, COLT (Desert Land, Reservoir and Canal), 1910, William Colt, 200,000 acres proposed north of La Junta in Otero and Bent Counties, $3 million proposed, no bonds issued, withdrawn by state land board 1913, no land reclaimed.

NO. 26, STARK-HAGADORN IRRIGATION, 1909, Edwin R. Stark and James D. Hagadorn, 28,000 acres proposed near Saguache in Saguache County, $400,000 proposed, no bonds issued, canceled by state land board 1920, no land reclaimed.

NO. 27, HAPPY HOME RESERVOIR, 1910, George Bergen, R. G. Troter, and H. W. Wright, 46,000 acres proposed near Ridgway in Montrose and Ouray Counties, $500,000 proposed, no bonds issued, canceled by state land board 1914, no land reclaimed.

NO. 28, BLUE MOUNTAIN IRRIGATION, 1910, James H. Clark, 177,000 acres proposed near Rangely in Rio Blanco and Moffat Counties, no proposed dollar outlay, no bonds issued, canceled by state land board 1914, no land reclaimed.

NO. 29, LAFAYETTE M. HUGHES, 1909, Lafayette M. Hughes, 11,200 acres proposed near Craig in Moffat County, no proposed dollar outlay, no bonds issued, canceled by state land board 1914, no land reclaimed.

NO. 30, WHITE RIVER, TRAPPERS LAKE RESERVOIR, 1909, various petitioners G. J. Magenheimer; E. Salisbury Smith and W. Rolla Wilson; C. G. Kirkland, W. J. Fine, H. D. Boughner, and D. S. Hamilton, 160,000 acres reduced to 91,400 acres proposed northeast of Meeker in Rio Blanco and Moffat Counties, $2 million proposed, no bonds issued, canceled by state land board 1914, no land reclaimed.

NO. 31, INTERSTATE IRRIGATION, 1911, Selma Miller and C. W. Bloodgood, 135,164 acres proposed near Durango in La Plata County, $4 million proposed, no bonds issued, canceled by state land board 1914, no land reclaimed.

NO. 32, KERBER CREEK, 1911, G. W. Gerard, 9,000 acres proposed near Villa Grove in Saguache County, no dollar outlay, no bonds issued, rejected by state land board 1912.

NO. 33, ROD MCDONALD, 1911, Rod McDonald, 30,000 acres proposed east of Trinidad in Las Animas County, no proposed dollar outlay, no bonds issued, canceled by state land board 1912, no land reclaimed.

NO. 34, TYRONE, 1911, C. L. Colburn, 36,000 acres proposed near Tyrone in Las Animas County, $1 million proposed, no bonds issued, canceled by state land board 1913, no land reclaimed.

NO. 35, ELK RIVER IRRIGATION AND CONSTRUCTION, 1911, Earle Wilkins, W. L. Jameson, and William Finley, 158,220 acres proposed north of Hayden in Routt and Moffat Counties, $1.3 million proposed, no bonds issued, canceled by state land board 1924, no land reclaimed.

NO. 36, PUEBLO AND NORTHEASTERN IRRIGATION, 1911, Kent Greenwald, 15,280 acres proposed south of Pueblo in Pueblo County, no dollar outlay, no bonds issued, canceled by state land board 1914, no land reclaimed.

NO. 37, DOLORES IRRIGATION, 1913, Frank E. Gove and George E. West, 196,650 acres proposed near Cortez in Montezuma County, $4 million proposed, no bonds issued, canceled by state land board 1914, no land reclaimed.

NO. 38, same as List No. 15 (Great Northern Irrigation and Power).

NO. 39, JACKSON COUNTY LAND AND IRRIGATION, 1913, no record.

NO. 40, added to List No. 18 (Duncan Chisholm), 1913, 2,240 acres.

NO. 41, added to List No. 18 (Duncan Chisholm), 1913, 17,900 acres.

NO. 42, DAVID H. SMITH, 1913, David H. Smith, 29,969 acres proposed near Craig in Moffat County, no dollar outlay, no bonds issued, canceled by state land board 1914, no land reclaimed.

NOTES

ABBREVIATIONS

CSA: Colorado State Archives
MFS: Map and Filing Statement
OCSE: Office of Colorado State Engineer
SBLC-CSA: State Board of Land Commissioners of Colorado–Colorado State Archives
TBM: Dr. Verity Two Buttes Museum

INTRODUCTION

1. Abbott, *Colorado*, 143–151.
2. Pisani, *From the Family Farm to Agribusiness*, 70–77.
3. On the Gem State, see Lovin, *Complexity in a Ditch*; Pisani, *Water and American Government*; Fiege, *Irrigated Eden*. Bonner, *William F. Cody's Wyoming Empire*, is the only monograph that examines the Carey Act in the Cowboy State. Walker, "The Delta Project," celebrates a noted Beehive State enterprise. On the historical importance of Colorado's soil conservation program, which effectively eradicated the Dust Bowl conditions, see Sheflin, *Legacies of Dust*. On the monumental significance of water conservancy policy, see Tyler, *Last Water Hole in the West*; Schulte, *As Precious as Blood.*

4. On the teleological argument about agriculture-to-conservation conversions, see Hamblin, "Philip Garone," 1–31.
5. Cronon, *Changes in the Land*; White, *"It's Your Misfortune and None of My Own"*; Worster, *Rivers of Empire*.
6. Tyler, *Last Water Hole in the West*; Weeks, *Cattle Beet Capital*.
7. Sherow, "Chimerical Vision."
8. Fiege, *Irrigated Eden*.
9. Garone, *Fall and Rise of the Wetlands of California's Great Central Valley*.
10. Orsi, "From Horicon to Hamburgers and Back Again."
11. Sherow, *Watering the Valley*; Sheflin, *Legacies of Dust*; Schulte, *As Precious as Blood*; Sturgeon, *Politics of Western Water*; Daniel Tyler, *Last Water Hole in the West*; Michael Weeks, *Cattle Beet Capital*; Stiller, *Water and Agriculture in Colorado and the American West*; Archer, *Unruly Waters*; Bogener, *Ditches across the Desert*; Logan, *Lessening Stream*.

CHAPTER 1: EXPERIMENT OR SWINDLE

1. *Cheyenne Daily Sun*, January 5, 1894, 1.
2. White, *"It's Your Misfortune and None of My Own,"* 85–118; Flores, *American Serengeti*, 1–9.
3. Larson, *History of Wyoming*, 69, 123, 141, 165. Carey quote is from the *Rocky Mountain News*, September 6, 1894, 8.
4. MacKinnon, *Public Waters*, 64, 67.
5. Pisani, *To Reclaim a Divided West*, 233–240.
6. Carey, "What the Carey Act Has Done for the West," in *Proceedings of the Conference of Western Governors Held at Salt Lake City, Utah June 5, 6, and 7, 1913*, 70–74.
7. Pisani, *To Reclaim a Divided West*, 230–232.
8. Pisani, *To Reclaim a Divided West*, 230–232; House, Cong. Rec., 50th Cong., 1st Sess., April 2, 1888, H.R. 9053, 2616; Pisani, *To Reclaim a Divided West*, 143–168.
9. Pisani, *To Reclaim a Divided West*, 230–232; Carey, "What the Carey Act Has Done for the West," in *Proceedings of the Conference of Western Governors Held at Salt Lake City, Utah June 5, 6, and 7, 1913*, 73; Cong. Rec., 51st Cong., 1st Sess., July 10, 1890, H.R. 11356, 7132.
10. Pisani, *To Reclaim a Divided West*, 233–234, 237–240.
11. Pisani, *To Reclaim a Divided West*, 235–237, 287–288. On Smythe's racism, see Limerick, *Desert Passages*, 77–90.
12. Pisani, *To Reclaim a Divided West*, 240–248.
13. Freeman, *Persistent Progressives*, 6.
14. Cong. Rec., 53rd Cong., 2d Sess., February 1, 1894, S. 1544, 1761, February 12, 1894, S. 1591, 2079, 4645; Report Amending and Favoring S. 1591, 53rd Cong., 2d Sess., April 17, 1894, Senate Report 332, Serial 3183.
15. Cong. Rec., 53rd Cong., 2d Sess., July 18, 1894, 9025–9026.
16. Willis Sweet bills H.R. 7152–7154 were substituted with H.R. 7558 (Cong. Rec., 53rd Cong., 2d Sess., May 21, 1894, 6215, 6042; Report Submitting H.R. 7558 for Survey of Reservoirs for Reclamation of Arid Lands, as Substitute for H.R. 7154, 53rd Cong., 2d Sess., June 23, 1894, House Report 1152, Serial 3272); *Aspen Weekly Times*, June 9, 1894,

3. On Thomas C. Power's bill, S. 2248, see Senate Journal, 53rd Cong., 2d Sess., July 23, 1894, 306; *Aspen Weekly Times*, July 28, 1894, 2.

17. Cong. Rec., 53rd Cong., 2d Sess., July 18, 1894, 9025–9026; Carey, "What the Carey Act Has Done for the West," 73.
18. Cong. Rec., 53rd Cong., 2d Sess., August 11, 1894, 9837, 9852–9853.
19. Cong. Rec., 53rd Cong., 2d Sess., August 11, 1894, 9735, 9856–9857, 9879–9888, 9966, 10075; Pisani, *To Reclaim a Divided West*, 402.
20. Cong. Rec., 53rd Cong., 2d Sess., August 11, 1894, 9963–9964.
21. Cong. Rec., 53rd Cong., 2d Sess., August 11, 1894, 9888.
22. Cong. Rec., 53rd Cong., 2d Sess., August 28, 1894, 10421–10426.
23. Cong. Rec., 53rd Cong., 2d Sess., August 28, 1894, 10421–10426.
24. US Statutes at Large, 28:422–423.
25. *Official Proceedings of the Third National Irrigation Congress*, 49; Pisani, *To Reclaim a Divided West*, 265, 408. The Carey quote is from the *Rocky Mountain News*, September 6, 1894, 8.
26. *Daily Boomerang* (Laramie, WY), September 7, 1894, 2; Larson, *History of Wyoming*, 304; Pisani, *To Reclaim a Divided West*, 252. Henry A. Coffeen, who was also an irrigator of some prominence in Wyoming, lost his bid for reelection in 1894, returned to Sheridan, and became something of a literary writer. Schulp, "I Am Not a Cuckoo Democrat." Mackey, *Henry A. Coffeen*, gives greater depth to his Populist views.
27. *Rocky Mountain News*, January 5, 1895, 4.
28. Laws of Colorado, 1895, ch. 70, An Act; *Summit County Journal*, February 23, 1895, 6; March 9, 1895, 6; *Boulder Camera*, March 12, 1895, 2.
29. *Leadville Evening Chronicle*, January 10, 1895, 2.
30. Larson, *History of Wyoming*, 303, 348–351; Bonner, "Elwood Mead"; Pisani, *To Reclaim a Divided West*, 398; Kluger, *Turning on Water with a Shovel*, 11.
31. Kluger, *Turning on Water with a Shovel*, 150; Hansen, *Democracy's College in the Centennial State*, 67–72; Teele, "Nettleton, Edwin S."
32. Hansen, *Democracy's College in the Centennial State*, 70; Worster, *Rivers of Empire*, 182–195.
33. Pisani, *To Reclaim a Divided West*, 261–264.
34. *Aspen Daily Times*, September 28, 1895, 3; December 7, 1895, 3; Bonner, *William F. Cody's Wyoming Empire*, is the full account of Cody's Carey Act involvement.
35. *Greeley Tribune*, December 24, 1896, 4; *Silverton* (CO) *Standard*, February 1, 1896, 4; *Golden* (CO) *Transcript*, April 1, 1896, 2; Bonner, *William F. Cody's Wyoming Empire*, 29, 33–36; Pisani, *To Reclaim an Arid West*, 261; Larson, *History of Wyoming*, 304.
36. *New York Times*, March 10, 1896, 2; US Statutes at Large 29:434, 31:1188.
37. Washington and California chose not to participate in the Carey Act. The other states accepting the entitlement were Arizona, which segregated 13,745 acres and patented no acres; Colorado, 284,654 and 37,302; New Mexico, 7,604 and 4,743; Nevada, 36,808 and 1,578; Utah, 141,814 and 37,239; Montana, 246,698 and 92,280; Oregon, 388,876 and 73,442; Wyoming, 1,396,869 and 203,311; and Idaho, 1,335,787 and 617,334. From Gates, *History of Public Land Law Development*, 651.

CHAPTER 2: CAREY ACT BEGINNINGS IN THE CENTENNIAL STATE

1. Underwood, *Town Building on the Colorado Frontier*; Raley, "Private Irrigation in Colorado's Grand Valley"; *Akron* (CO) *Weekly Pioneer Press*, September 13, 1895, 1.
2. Pisani, *To Reclaim a Divided West*, 57–58, 208–210, 212; Dunbar, *Forging New Rights in Western Waters*, 24–26, 101.
3. *Akron* (CO) *Weekly Pioneer Press*, September 13, 1895, 1: Dille, *Irrigation in Morgan County*; Tyler, *Last Water Hole in the West*, 47; Stone, *History of Colorado*, vol. 2, 178–179.
4. Whol, *Wide Rivers Crossed*, 33–45, 63–104.
5. Whol, *Wide Rivers Crossed*, 49–55.
6. Office of Colorado State Engineer, *Biennial Report, 1895–1896*, 39; *Castle Rock* (CO) *Journal*, March 4, 1896, 2; *Great West* 1, 1 (July 1896), 26–27; *Greeley Tribune*, December 24, 1896, 3; Hinton, *Irrigation in the United States*, 127, 129.
7. *Fort Collins Courier*, February 18, 1897, 2; Pisani, *To Reclaim a Divided West*, 270, 410–411; *Irrigation Age* 11, 2 (February 1897), 21–22; *Chafee County* (CO) *Republican*, February 24, 1897, 4; US Geological Survey, *Annual Report, 1902–1903*, 225–226.
8. Riverside Reservoir and Land Company v. Bijou Irrigation District et. al.; US Reclamation Service, *Annual Report, 1902–1903*, 154–160; *Irrigation Age* 17, 10 (October 1902, 372–377; US Reclamation Service, *Annual Report, 1902–1903*, 162–182.
9. Tyler, *Silver Fox of the Rockies*, 77–136; Tyler, *Last Water Hole in the West.*
10. Colorado Enabling Act, March 3, 1875, Sec. 7–10, 14; State Board of Land Commissioners of Colorado, *Biennial Report, 1905–1906*, 3–4.
11. Laws of Colorado, 1879, An Act, 171–174; 1881, An Act, 223–226; 1887, An Act, 328–338; First Report of the Public Examiner of Colorado, 1909–1910, 70–71.
12. Chamberlain, *Principals of Bond Investment*, 384–400; Teele, "Financing of Non-Governmental Irrigation Enterprises," 427–428, 430–431, 434–435; Lovin, "LaSalle Street Capitalists."
13. Teele, "Financing of Non-Governmental Irrigation Enterprises," 427–435.
14. Teele, "Financing of Non-Governmental Irrigation Enterprises," 427–435.
15. Worster, *Rivers of Empire.*
16. Simmons, *Ute Indians of Utah, Colorado, and New Mexico.*
17. Stone, *History of Colorado*, vol. 4, 804–805; *Silver Cliff* (CO) *Rustler*, June 25, 1902, 3; Land Board Minutes Concerning, Desert Lands, Container 12571G, 1900–1909, List 3, 5–31, 95, 526–529, State Board of Land Commissioners of Colorado, Colorado State Archives (hereafter SBLC-CSA); *Sugar Beet Gazette* 6, 5 (March 5, 1904), 99; *Steamboat Pilot*, August 5, 1903, 1; Routt County Development Company v. Johnston et al.; Carey Act and Desert Land Records, Container 12571D, 1902–1914, Snake River Canal, SBLC-CSA.
18. Goldberg, *Hooded Empire*, 22; *Fort Collins Courier*, August 7, 1907, 11; *Routt County* (CO) *Sentinel*, August 2, 1907, 1; *Steamboat Pilot*, November 20, 1907, 1; May 6, 1908, 1; *Routt County* (CO) *Sentinel*, July 9, 1909, 1; May 27, 1910, 1; *Steamboat Pilot*, November 3, 1908, 1; Land Board Minutes Concerning, Desert Lands, container 12571G, 1900–1909, List 3, 5–31, 95, 190–232, SBLC-CSA; State Engineer of Colorado, *Biennial Report, 1909–1910*, 13; *Seventeenth Annual Report of the Commission of Banking*, 711; *Craig* (CO) *Daily Press*,

March 14, 2009, accessed November 3, 2020, www.craigdailypress.com/news/2009/mar/14/wantland; *Steamboat Pilot*, May 3, 1911, 6; *Oak Creek* (CO) *Times*, April 27, 1916, 1.

19. Mehls, "An Area the Size of Pennsylvania"; Mehls, "Westward from Denver"; *Sugar Beet Gazette* 6, 5 (March 5, 1904), 99; US Reclamation Service, *Annual Report, 1902–1903*, 206–209; Carey Act and Desert Land Records, Container 12571D, 1902–1914, Great Northern Irrigation and Power, SBLC-CSA; Carey Act Lands and Indemnity and Lands Docket, Container 12571L, 1902–1911, SBLC-CSA; *Steamboat Pilot*, February 1, 1905, 1.
20. Map and Filing Statement (hereafter MFS) 5360, Great Northern Irrigation System, Dist. 44, Office of Colorado State Engineer (hereafter OCSE); *Steamboat Pilot*, February 1, 1905, 1; Carey Act and Desert Land Records, Container 12571D, 1902–1914, Great Northern Irrigation and Power, SBLC-CSA; *Breckenridge* (CO) *Bulletin*, July 13, 1907, 2; *Routt County* (CO) *Republican*, June 18, 1909, 1; Land Board Minutes Concerning, Desert Lands, Container 12571G, 1900–1919, List 15, 526–529, SBLC-CSA; State Engineer of Colorado, *Biennial Report, 1909–1910*, 12; *Steamboat Pilot*, November 17, 1909; Wayne Wymore Collection, 04–27, Museum of Northwest Colorado; *Routt County* (CO) *Republican*, January 2, 1914, 1; Monthly Compendium, 66th Cong., Complete, No. 16, March 1921, 175; US Statutes at Large, 41:407; *Relief to Settlers under Forfeited Carey Act Projects*, 65th Cong., 2d Sess., January 25, 1918; *Relief to Settlers under Forfeited Carey Act Projects*, 65th Cong., 3d Sess., February 6, 1919; *Preference Right of Entry by Certain Carey Act Entrymen*, 66th Cong., 1st Sess., August 2, 1919; *Preference Right of Entry by Certain Carey Act Entrymen*, 66th Cong., 2d Sess., January 8, 1920.
21. Laws of Colorado, 1907, Senate Joint Memorial No. 2; *Durango* (CO) *Democrat*, February 15, 1907, 4; Grant of Certain Lands to Colorado, 59th Cong., 2d Sess., February 25, 1907, Senate R. 7279, Serial 5061, 1–3; US Statutes at Large, 34:1056.
22. *Directory of Colorado Springs*, 193; United States v. S. W. Morrison et al., discussion in US Department of the Interior, *Annual Report, 1901*, Indian Affairs, I, 623–624; Statement of Claim, Dr. Morrison Ditch, No. CA1248, ID 505 (000162548), Dist. 31, Water Rights, OCSE. Also see Quintana and Clemmer, 35–38; Land Board Minutes Concerning, Desert Lands, Container 12571G, 1900–1909, List 6, 100–135, SBLC-CSA; US Statutes at Large, 31:479, 31:179; 31:180, 31:740; *Durango* (CO) *Democrat*, February 15, 1907, 4; May 12, 1909, 2; July 17, 1909, 1; Fine-Dare, *Grave Injustice*, 58–61 (Wadleigh quote); *Durango* (CO) *Wage Earner*, October 12, 1911, 1; *Bayfield* (CO) *Blade*, June 30, 1910, 1; July 14, 1910, 1; July 28, 1910, 1; November 2, 1911, 1; MFS 6380, *Ignacio Ditch* (Enlargement), Dist. 31, OCSE.
23. Carey Act and Desert Land Records, Container 12571D, 1902–1914, Toltec Canal Company, SBLC-CSA; MFS 34, Taos Valley Irrigation Canal and Reservoir System, Dist. 22, OCSE; *Creede* (CO) *Candle*, October 23, 1909, 3; October 30, 1909, 4; MFS 2320, Cove Lake Reservoir, Dist. 22, OCSE; Historical Diversion Records–Commissioner's Report, Water Dist. 22, Taos Valley Canal 1–3, 1906, 1909, OCSE; Land Board Minutes Concerning Desert Lands, Container 12571G, 1900–1909, List 10, 464–486, SBLC-CSA.

24. Stiller, *Water and Agriculture in Colorado and the American West*; Raley, "Private Irrigation in Colorado's Grand Valley"; Sturgeon, *Politics of Western Water*; Schulte, *Wayne Aspinall*; Schulte, *As Precious as Blood*; Harvey, *Symbol of Wilderness*.
25. Newell, *Report on Agriculture by Irrigation in the Western Part of the United States at the Eleventh Census*, 102.

CHAPTER 3: THE COLONY AT TWO BUTTES MOUNTAIN

1. Colorado Millennial Site/Hackberry Site/Bloody Springs, History Colorado; Grinnell, *Fighting Cheyennes*; Hämäläinen, *Comanche Empire*.
2. MFS 2665, February 16, 1906, Wm. D. Purse Reservoir No.1 Dist. 67, OCSE; MFS 2666, March 14, 1906, Wm. D. Purse Reservoir No. 2, Dist. 67, OCSE; Two Buttes Irrigation and Reservoir Company Right of Way Application, Serial No. 05275, 1906–1908, US Bureau of Land Management. On the perception of nature as a factor of development, see Logan, *The Lessening Stream*.
3. MFS 89, 1902, Purse and Mador Ditch, District 17, OCSE.
4. *Twelfth Annual Report of the Director of the United States Geological Survey [1890–1891], Part 2*, "Irrigation," 106–107; Pisani, *To Reclaim a Divided West*, 143–168; Brookings Institution for Government Research, *U.S. Reclamation Service*, 13.
5. Colorado State Planning Division, *Year Book of the State of Colorado, 1959–1961* (Denver, 1961), 827.
6. Hill, "History of Baca County," 112–117, 215; Taylor, "Town Boom in Las Animas and Baca Counties."
7. MFS 3917, July 30, 1899, Butte Creek Ditches Nos. 1 and 2, Dist. 67, OCSE; MFS 134, May 25, 1903, Felix Cain Ditch, Dist. 67, OCSE; OCSE; Osteen, *A Place Called Baca*, 78; Hargreaves, "Hardy Webster Campbell (1850–1937)"; Hargreaves, *Dry Farming in the Northern Great Plains*, 85–99, 162; Webb, *The Great Plains*, 322, 366–374.
8. *Proceedings of the Trans-Missouri Dry Farming Congress*, 108–111, 37–43.
9. *Proceedings of the Trans-Missouri Dry Farming Congress*, 39.
10. MFS 2665, February 16, 1906, Wm. D. Purse Reservoir No. 1, Dist. 67; MFS 2666, March 14, 1906, Wm. D. Purse Reservoir No. 2, Dist. 67; MFS 5000, February 16, 1906, Two Buttes Reservoir and Outlet Ditches, Dist. 67, OCSE.
11. Land Board Minutes Concerning Desert Lands, Container 12571G, 1900–1909, List 7, 328–385, SBLC-CSA; Carey Lands and Indemnity Lands Docket, Container 12571L, 1902–1911, SBLC-CSA.
12. *Lamar* (CO) *Daily News*, January 4, 1941, 1–2; *Oregonian*, March 14, 1911, 6; Idaho State Historical Society, Twin Falls Land and Water Company Files, MS114 General Correspondence (Secretary's Files); *Official Proceedings of the Fourteenth National Irrigation Congress*, 228, 236; *Commercial and Financial Chronicle* 84, Finance and Investments Section (January 12, 1907), 108; *Moody's Magazine* 5 (June 1908), 2; *Denver Post*, January 26, 1931, 4.
13. Land Board Minutes Concerning Desert Lands, Container 12571G, 1900–1909, List 7, 328–340, SBLC-CSA; William D. Purse to Thomas W. Jaycox, August 1908,

September 9, 1908, in MFS 4895, February 16, 1906, Two Buttes Reservoir and Outlet Ditches, Dist. 67, OCSE; Articles of Incorporation, Two Buttes Reservoir and Irrigation Company, August 17, 1908, Dr. Verity Two Buttes Museum, Two Buttes, CO (hereafter TBM).

14. William D. Purse to Thomas W. Jaycox, August 1908, September 9, 1908; Lincoln Bancroft to Thomas W. Jaycox, October 7, 1908, all in MFS 4895, February 6, 1906, Two Buttes Reservoir and Outlet Ditches, Dist. 67, OCSE.
15. Granby Hillyer to Thomas W. Jaycox, November 16, 1908, Thomas W. Jaycox to Granby Hillyer, November 18, 1908, Granby Hillyer to Thomas W. Jaycox, December 11, 1908, Thomas W. Jaycox to John F. Vivian, December 15, 1908, all in MFS 4895, February 6, 1906, Two Buttes Reservoir and Outlet Ditches, Dist. 67, OCSE.
16. Land Board Minutes Concerning Desert Lands, Container 12571G, List 7, 1900–1909, 328–340, SBLC-CSA; Carey Lands and Indemnity Lands Docket, Container 12571L, 1902–1911, SBLC-CSA.
17. Carey Act and Desert Land Records, Container 12571D, 1902–1914, SBLC-CSA.
18. Carey Act and Desert Land Records, Container 12571L, 1902–1911, SBLC-CSA; MFS 5000, Two Buttes Reservoir and Outlet Ditches, Dist. 67, OCSE; Land Board Minutes Concerning Desert Lands, Container 12571G, 1900–1909, List 7, 328–385, SBLC-CSA.
19. On Homer W. McCoy and Company and Municipal Securities Company investment in the Two Buttes Irrigation and Reservoir Company, see Fred L. Harris to McCoy and Company, March 2, 1911, TBM.
20. Land Board Minutes Concerning Desert Lands, Container 12571G, 1900–1909, List 7, 328–385, SBLC-CSA.
21. *Colorado Carey Act Land Opening*, photocopy in possession of TBM.
22. *Colorado Carey Act Land Opening*, photocopy in possession of TBM.
23. Desert Land Selections Reception Book, Container 12571H, List 7, 1907–1911, SBLC-CSA; *Colorado Transcript* (Golden), October 7, 1909, 6; Baca County Historical Society, *Baca County*, 81; Harper, "Development of a High Plains Community," 42–43.
24. Water Contract Ledger Book, TBM. Harris, *Long Vistas*, makes the point that women were active participants in settling government-gifted land. On the larger role of women in the West, see Deutsch, *Making a Modern West*.
25. Land Board Minutes Concerning Desert Lands, Container 12571G, List 7, 1900–1909, 328–385, SBLC-CSA.
26. Resolutions 1 and 2, Board of Directors Two Buttes Irrigation and Reservoir Company, November 26, 1910, TBM; Fred L. Harris to McCoy and Company, March 2, 1911, TBM; Teele, "Financing of Non-Governmental Irrigation Enterprises," 433.
27. State Engineer of Colorado, *Biennial Report, 1909–1910*, 125–126; *Irrigation Age* 25, 1 (November 1909), 28; Baca County Historical Society, *Baca County*, 84–85.
28. State Engineer of Colorado, *Biennial Report, 1909–1910*, 125–126.
29. Baca County Historical Society, *Baca County*, 84–85; Carey Lands and Indemnity Lands Docket, Container 12571L, 1902–1911, SBLC-CSA.
30. Carey Lands and Indemnity Lands Docket, Container 12571L, 1902–1911, SBLC-CSA.
31. Desert Lands and Carey Lands Minutes, Container 12571G, List 7, 1900–1909, 328–385, SBLC-CSA; Keating, *The Gentleman from Colorado*, 131–133; Desert Lands and

Carey Lands Minutes, Container 12571B, List 7, 1912–1914, 16–18, SBLC-CSA, contains 1911 records; *Steamboat Pilot*, September 13, 1911, 8.

32. Fred L. Harris, *Yearly Precipitation and Reservoir Record Book*, TBM.
33. *Poor's Manual of Industrials, Manufacturing, Mining and Miscellaneous Companies*, 3007–3008.
34. Carey Act Projects, 62d Cong., 3d Sess., February 21, 1913, S. Doc. 1097, Serial 6365, 3–22; Charles W. Wells to Commissioner, October 29, 1912, Lamar Serial No. 01190, Box 866, Letters Received, Records of the General Land Office, RG 49, National Archives, Denver.
35. Desert Lands and Carey Lands Minutes, Container 12571B, List 7, 1912–1914, 57, SBLC-CSA; *Fairplay* (CO) *Flume*, September 26, 1913, 2; *Denver Farm and Field*, September 6, 1913, 6; September 27, 6.
36. State Engineer of Colorado, *Biennial Report, 1913–1914*, 210–218.
37. Fred L. Harris to William D. Purse, June 1, 1914, June 8, 1916, TBM.

CHAPTER 4: THE BIRTH OF THE MUDDY CREEK CAREY ACT PROJECT

1. Scott, *Geologic and Structure Contour Map of the La Junta Quadrangle, Colorado and Kansas*; Long, "Basal Cretaceous Strata"; *Twelfth Annual Report of the Director of the United States Geological Survey [1890–1891]*, Part 2, "Irrigation," 101–104.
2. MFS 99, December 1, 1902, Chaquaqua Irrigation System, Dist. 17, OCSE. The Bent and Prowers Irrigation District proposed watering farmlands located south of Lamar. Its organizers were unable to place its financing bonds, and the district was eventually abandoned after 1920. See "Bent and Prowers Irrigation District, Report," Lamar, 1912, TBM.
3. Stiller, *Water and Agriculture in Colorado and the American West*, 3–10; Pisani, "Enterprise and Equity," 30.
4. Clark, *On the Edge of Purgatory*.
5. Carrillo, "In-Depth Review of Regional History"; Stone, *History of Colorado*, vol. 2, 411–412. For examples of the typical racial tropes, see *Las Animas* (CO) *Leader*, March 28, 1902, 2; February 2, 1906, 4.
6. Carrillo, "In-Depth Review of Regional History."
7. Stone, *History of Colorado*, vol. 2, 411–412.
8. Murphy, "History of Carey Act Selection No. 11," 8–12.
9. Murphy, "History of Carey Act Selection No. 11," 8–12.
10. Murphy and others, *Destructive Floods in the United States in 1904*, 158–169; MFS 3340, 1907, Smith Canyon Reservoir, Dist. 19, OCSE; MFS 4036, 1907, Muddy Creek Reservoir, Dist. 67, OCSE; Frank H. Whiting, *Maps Showing the Drainage Area of the Las Animas–Bent County Irrigation Project* (Denver, 1909), Carey Act No. 11 File, Land Records, Bureau of Land Management, US Department of the Interior, Lakewood, CO; Sherow, *Watering the Valley*, 36–37.

11. Sherow, "The Chimerical Vision"; Hicks, "Storage of Storm-Waters on the Great Plains"; Mahard, "History of the Department of Geology and Geography," 19.
12. Pisani, "Reclamation and Social Engineering in the Progressive Era"; Sherow, "The Chimerical Vision"; Stone, *History of Colorado*, vol. 2, 411–412.
13. Sherow, *Watering the Valley*, 103; Beach and Preston, *Irrigation in Colorado*, 18.
14. Stone, *History of Colorado*, vol. 2, 411–412; *San Juan* (CO) *Prospector*, October 11, 1918, 2; *Haswell* (CO) *Herald*, October 24, 1918, 8.
15. The Valley Investment Company Right of Way Application, Serial No. 011696, Bureau of Land Management, US Department of the Interior, Lakewood, CO; *Portrait and Biographical Record of Dubuque, Jones and Clayton Counties, Iowa*, 260–261; *Steamboat Pilot*, November 29, 1905, 1; *Yampa* (CO) *Leader*, July 13, 1907, 8; *Steamboat Pilot*, December 22, 1909, 3; US Department of Commerce, 1900 Census, Colorado, Arapahoe, Series T623, Roll 118, 297; US Department of Commerce, 1910 Census, Colorado, Denver, Series T624, Roll 115, 112.
16. Murphy, "History of Carey Act Selection No. 11," 8–12.
17. Frank H. Whiting Photograph Collection, Alaska State Library; Murphy, "History of Carey Act Selection No. 11," 8–12; *Biennial Report of the Secretary of State of Colorado, 1907–1908*, 50.
18. Stone, *History of Colorado*, vol. 2, 411–412; Wieber, "History of the United States Naval Hospital, Fort Lyon, Colo.," 745–747; "Register of Funerals, 1896–1910."
19. Freeman, Lamb, and Bolster, *Surface Water Supply of the United States, 1907–1908*, 31, 33–40.
20. Freeman, Lamb, and Bolster, *Surface Water Supply of the United States, 1907–1908*, 31, 33–40.
21. Freeman, Lamb, and Bolster, *Surface Water Supply of the United States, 1907–1908*, 31, 33–40, quotation p. 40.
22. Carey Act and Desert Land Records, Container 12571D, 1902–1914, Valley Investment, SBLC-CSA; Carey Lands and Indemnity Lands Docket, Container 12571L, 1902–1911, SBLC-CSA.
23. Whiting, *Maps Showing the Drainage Area of the Las Animas–Bent County Irrigation Project*.
24. Murphy, "History of Carey Act Selection No. 11," 8, 40 (Comstock quote); Carey Act and Desert Land Records, Container 12571D, 1902–1914, Valley Investment, SBLC-CSA.
25. MFS 2692, 1907, Bent County Reservoir, Dist. 67, OCSE; *Irrigation Age* 25, 3 (January 1910), 129–130.
26. On the Doctrine of Appropriation relative to Colorado, see Schorr, *Colorado Doctrine*. On Kansas v. Colorado, see Sherow, *Watering the Valley*.
27. Charles W. Wells to Commissioner, General Land Office, June 18, 1910, Carey Act No. 11 File, Land Records, Bureau of Land Management, US Department of the Interior, Lakewood, CO; Carey Act and Desert Land Records, Container 12571D, 1902–1914, Valley Investment, SBLC-CSA.
28. Gerald G. O'Brien, interview with the author, June 17, 1976; *Official Proceedings of the Eighteenth National Irrigation Congress*.

29. US Statutes at Large, 36:835. From this $20 million loan, the Reclamation Service authorized $1.5 million and $1 million, respectively, to its Colorado projects: the Uncompahgre and Grand Valley developments. See US Department of the Interior, *Bureau of Reclamation Project Feasibilities and Authorizations*, 20–21.
30. Pisani, *Water, Land, and Law in the West*, 66–67, 323; Pisani, *Water and American Government*, 104–113; Lovin, "Farwell Trust Company of Chicago."
31. State Engineer of Colorado, *Biennial Report, 1915–1916*, 7.
32. Whiting, *Maps Showing the Drainage Area of the Las Animas–Bent County Irrigation Project*; Brookings Institution for Government Research, *U.S. Reclamation Service*, 5–8; Charles W. Wells to Commissioner, October 5, 1912; Murphy, "History of Carey Act Selection No. 11," 9.

CHAPTER 5: "LAND BOARD, MAKE THEM FISH OR CUT BAIT"

1. Additional Lands for Colorado under Provisions of the Carey Act, 62d Cong., 1st Sess., August 4, 1911, H. Report 120, Serial 6078, and July 26, 1911, S. Report 110, Serial 6077.
2. Additional Lands for Colorado under Provisions of the Carey Act, 62d Cong., 1st Sess., August 4, 1911, H. Report 120, Serial 6078, and July 26, 1911, S. Report 110, Serial 6077; *Eagle Valley* (CO) *Enterprise*, August 25, 1911, 4; US Statutes at Large 37:38.
3. Additional Lands for Colorado under Provisions of the Carey Act, 62d Cong., 1st Sess., August 4, 1911, H. Report 120, Serial 6078.
4. Laws of Colorado, 1909, Chapter 149.
5. State Board of Land Commissioners of Colorado, *Biennial Report, 1909–1910*, 137–142; *Steamboat Pilot*, January 18, 1911, 3.
6. Laws of Colorado, 1909, House Joint Resolution No. 6, Land Board Investigation; *Aspen Democrat-Times*, October 9, 1909, 1.
7. State Board of Land Commissioners of Colorado, *Biennial Report, 1909–1910*, 3–4; *State Engineer of Colorado, Biennial Report, 1909–1910*, 26.
8. *Aspen Daily Times*, January 31, 1909, 1; *Oak Creek* (CO) *Times*, September 9, 1922, 8.
9. State Board of Land Commissioners of Colorado, *Biennial Report, 1909–1910*, 3.
10. State Board of Land Commissioners of Colorado, *Biennial Report, 1909–1910*, 3–10; *Oak Creek* (CO) *Times*, September 22, 1910, 1; *Steamboat Pilot*, January 18, 1911, 3.
11. *Routt County* (CO) *Republican*, July 28, 1911, 4; State Board of Land Commissioners of Colorado, *Biennial Report, 1909–1910*, 6–7.
12. State Board of Land Commissioners of Colorado, *Biennial Report, 1919–1920*, 88–89.
13. *Routt County* (CO) *Republican*, August 18, 1911, 4.
14. Carey Act Projects, 62d Cong., 3d Sess., February 21, 1913, S. Doc. 1097, Serial 6365, 3–22; Wells to Commissioner, October 29, 1912, Lamar Serial No. 01190, and October 5, 1912 (source of quotations), Lamar Serial No. 06681, Box 866, Letters Received, Records of the General Land Office, Record Group (hereafter RG) 49, National Archives, Denver. The October 5, 1912, letter that contains the quotations cited is

duplicated in the file of Carey Act No. 11, Bureau of Land Management, US Department of the Interior, Lakewood, CO.

15. Carey Act Projects, 62d Cong., 3d Sess., February 21, 1913, S. Doc. 1097, Serial 6365, 5.
16. Carey Act Projects, 62d Cong., 3d Sess., February 21, 1913, S. Doc. 1097, Serial 6365, 4–5.
17. *Montezuma* (CO) *Journal*, April 21, 1913, 1; State Board of Land Commissioners of Colorado, *Biennial Report, 1911–1912*, 3–4, 1913–1914, 5–6; *Steamboat Pilot*, June 18, 1913, 1; Healy, "Admiral William B. Caperton," 84–88.
18. In re Salaries of Commissioners and Employees of State Land Board; *Weekly Ignacio* (CO) *Chieftain*, April 30, 1915, 2.
19. *Routt County* (CO) *Republican*, August 8, 1913, 4; August 11, 1913, 1; *Fort Collins Weekly Courier*, August 15, 1913, 7; *Record Journal of Douglas County* (CO), August 29, 1913, 4; Burroughs, *Where the Old West Stayed Young*, 321–330; State Board of Land Commissioners of Colorado, *Biennial Report, 1913–1914*, 3–4; *Routt County* (CO) *Republican*, August 22, 1913, 1; Woeste, *Henry Ford's War on Jews*, 153–154.
20. State Board of Land Commissioners of Colorado, *Biennial Report, 1913–1914*, 3–8; *Telluride Journal*, February 19, 1914, 3.
21. Desert Lands and Carey Lands Minutes, Container 12571B, 1912–1914, Canal Board Record, SBLC-CSA.
22. *Forestry and Irrigation* 10, 10 (October 1904), 447.
23. *Brandon* (CO) *Bell*, October 3, 1913, 3.
24. *Routt County* (CO) *Republican*, January 2, 1914, 1.
25. *Twelfth Annual Report of the Director of the United States Geological Survey [1890–1891], Part* 2, "Irrigation" (Washington, DC, 1892), 88–89; MFS 2508, 1906, Badito Reservoir, Dist. 16, OCSE; Frederic Auten Combs Perrine, "Report on the Power Development of the Huerfano River at Badito, Huerfano County, Colorado, 1907," CMSS-M28, Denver Public Library, Western History Collection, 1–57; MFS 5372, 1908, Badito Canals, S. and N. Lateral and Pipeline, Dist. 16, OCSE; *Brick* 25, 3 (September 1906), 138; *Electrical World* 49, 2 (January 12, 1907), 123; Carey Act and Desert Land Records, Container 12571D, 1902–1914, Huerfano Valley Irrigation, SBLC-CSA; State Engineer of Colorado, *Biennial Report, 1909–1910*, 13.
26. Ketchum, "Milo Smith Ketchum—Dean"; *Science* 32, 821 (September 23, 1910), 404; "Report of Crocker and Ketchum," in John Philip Donovan, "Report on Badito Irrigation Project," Donovan Papers, CMSS-M27, Western History Collection, Denver Public Library; State Engineer of Colorado, *Biennial Report, 1909–1910*, 13 (quotation).
27. Carey Act and Desert Land Records, Container 12571D, 1902–1914, Huerfano Valley Irrigation, SBLC-CSA.
28. Carey Act and Desert Land Records, Container 12571D, 1902–1914, Huerfano Valley Irrigation, SBLC-CSA; John Philip Donovan to W. L. Rucker [1912], in Donovan, "Report on Badito Irrigation Project."
29. Carey Act and Desert Land Records, Container 12571D, 1902–1914, Huerfano Valley Irrigation, SBLC-CSA; John Philip Donovan to W. L. Rucker [1912] and "Report of W. B. Freeman" n.d., both in Donovan, "Report on Badito Irrigation Project"; US Reclamation Service, *Annual Report, 1924*, 82–83.

30. Desert Lands and Carey Lands Minutes, Container 12571B, 1912–1914, Investigations Docket, 51–67, SBLC-CSA.
31. Desert Lands and Carey Lands Minutes, Container 12571B, 1912–1914, Investigations Docket, 51–67, SBLC-CSA.
32. Desert Lands and Carey Lands Minutes, Container 12571B, 1912–1914, Investigations Docket, 51–67, SBLC-CSA; *Weekly Ignacio* (CO) *Chieftain*, May 8, 1914, 1.
33. State Board of Land Commissioners of Colorado, *Biennial Report, 1913–1914*, 11–15.
34. State Board of Land Commissioners of Colorado, *Biennial Report, 1913–1914*, 11–15, quotation on p. 12.

CHAPTER 6: PERHAPS NOBODY IS TO BLAME

1. *Wyoming Weekly Tribune* (Cheyenne), April 7, 1914, 2; *Denver Post*, April 7, 1914, 1–2.
2. Larson, *History of Wyoming*, 290–292, 319–324.
3. Speech and quotations from *Proceedings of the Conference of Western Governors Held at Salt Lake City*, 74–77; Carey Act Projects, 62d Cong., 3d sess., 1913, S. Doc. 1097, Serial 6365, 17–20.
4. *Proceedings of the Conference of Western Governors Held at Denver*; US General Land Office, *Annual Report, 1914*, 3.
5. *Proceedings of the Conference of Western Governors Held at Denver*, 12.
6. *Proceedings of the Conference of Western Governors Held at Denver*, 12–17.
7. *Proceedings of the Conference of Western Governors Held at Denver*, 18–20.
8. *Proceedings of the Conference of Western Governors Held at Denver*, 20–24.
9. *Proceedings of the Conference of Western Governors Held at Denver*, 20–24.
10. *Akron* (CO) *Weekly Pioneer Press*, April 17, 1914, 8; *Montezuma* (CO) *Journal*, April 16, 1914, 2; *Irrigation Age* 29, 7 (May 1914), 201.
11. *Irrigation Age* 29, 7 (May 1914), 201.
12. Pisani, *To Reclaim a Divided West*, 263–264, 407–408.
13. Kluger, *Turning on Water with a Shovel*, 57–73.
14. Mead, "Systematic Aid to Settlers Is First Need," 202.
15. Pisani, *To Reclaim a Divided West*, 306–319; Jackson, "Engineering in the Progressive Era," 539–574.
16. Mead, "Systematic Aid to Settlers Is First Need," 202–203.
17. Mead, "Systematic Aid to Settlers Is First Need," 202–203.
18. Mead, "Systematic Aid to Settlers Is First Need," 204.
19. Mead, "Systematic Aid to Settlers Is First Need," 204, 216.
20. Kluger, *Turning on Water with a Shovel*, 57–73.
21. Teisch, *Engineering Nature*, 82–87; Davison, "Country Life."
22. Kluger, *Turning on Water with a Shovel*, 150–159.
23. Kluger, *Turning on Water with a Shovel*, 65–67: US General Land Office, *Annual Report, 1914*, 29.

CHAPTER 7: GEORGE E. O'BRIEN AND THE DISASTROUS MUDDY CREEK COLONY

1. Gerald G. O'Brien, interview with the author, June 11, 1982.
2. Land Board Minutes Concerning Desert Lands, Container 12571B, List 11, SBLC-CSA, 57; Murphy, "History of Carey Act Selection No. 11," 9–10, SBLC-CSA.
3. Hutchins, "Mutual Irrigation Companies"; Cech, *Principles of Water Resources*, 346–348. Worster, *Rivers of Empire*, 83–96, makes the exploitive argument and Schorr, *Colorado Doctrine*, the practical argument. Also see Weeks, *Cattle Beet Capital*, 31–33, 128–135.
4. Gerald G. O'Brien, interview with the author, June 11, 1982.
5. Declaration of Applicant, Desert Land Entry No. 08132, May 31, 1910; Application for Extension of Time, Desert Land Entry No. 08132, December 4, 1914; Application for Relief under Third Paragraph, Act of March 4, 1915 (Public 296), March 22, 1917; and Final Certificate, No. 08132, May 16, 1921, all in Serial Patent Number File 814755, United States National Archives and Records Administration, Washington, DC.
6. Gerald G. O'Brien, interview with the author, June 11, 1982.
7. Gerald G. O'Brien, interview with the author, June 11, 1982; McMahon Audit Company, "Financial Statement the Mutual Carey Irrigation Company," in "Reports on the Carey Act Selection No. 11, Bent County Colorado," 30–33, SBLC-CSA.
8. Murphy, "History of Carey Act Selection No. 11," 10, SBLC-CSA.
9. Hoggatt, *How to Get a Homestead in the National Forest Reserves*; *San Juan* (CO) *Prospector*, October 11, 1918, 2; *Fort Collins Courier*, March 22, 1919, 1; Attorney General of Colorado, *Biennial Report, 1919–1920*, 15; Burroughs, *Where the Old West Stayed Young*, 330.
10. *Fort Collins Courier*, March 22, 1919, 2; *Range Leader* (Hugo, CO), June 7, 1919, 4; Attorney General of Colorado, *Biennial Report, 1919–1920*, 15; People ex rel. Murphy v. Field, 181.
11. Murphy, "History of Carey Act Selection No. 11," 10, SBLC-CSA; Schrontz, "Report of the Cost of Completed Construction and Costs of Completion," SBLC-CSA.
12. Murphy, "History of Carey Act Selection No. 11," 10, SBLC-CSA; Schrontz, "Report of the Cost of Completed Construction and Costs of Completion," SBLC-CSA.
13. Murphy, "History of Carey Act Selection No. 11," 10, SBLC-CSA; Schrontz, "Report of the Cost of Completed Construction and Costs of Completion," SBLC-CSA, 16, 33.
14. O'Brien, "Spouse and Me," no date, Gerald C. Morton Collection.
15. O'Brien, "Spouse and Me."
16. O'Brien, "Spouse and Me."
17. O'Brien, "Spouse and Me."
18. O'Brien, "Spouse and Me."
19. "Financial Statement, the Mutual Carey Irrigation Company," January 31, 1922, in Murphy, "History of Carey Act Selection No. 11," 27–33.
20. Bray, *Financing Western Cattlemen*, 44; Wheelock, "Regulation and Bank Failures,"; Fite, *George N. Peek and the Fight for Farm Parity*, 4–20.
21. Carey Act and Desert Land Records, Elk River Irrigation, Great Northern Irrigation and Power, and Huerfano Valley Irrigation, Container 12571D, SBLC-CSA.
22. *Reports on Carey Act Selection No. 11, Bent County, Colorado*, 3–4, SBLC-CSA.

23. Benj. Griffith, *Report of the Colorado Irrigation District Finance Commission to the Twenty-Third General Assembly [of Colorado, 1921]*, 3–9.
24. Benj. Griffith, *Report of the Colorado Irrigation District Finance Commission to the Twenty-Third General Assembly [of Colorado, 1921]*, 44–45.
25. State Board of Land Commissioners of Colorado, *Biennial Report, 1921–1922*, 7–8; "Statement Concerning Carey Act Selection No. 11, Bent County, Colorado," in *Reports on Carey Act Selection No. 11, Bent County, Colorado*, 5–6, SBLC-CSA.
26. "Statement Concerning Carey Act Selection No. 11, Bent County, Colorado," in *Reports on Carey Act Selection No. 11, Bent County, Colorado*, 3–4, SBLC-CSA.
27. *Fort Collins Courier*, May 3, 1922, 11; Patent of Carey Segregation No. 11 to the State of Colorado, 67th Cong., 2d Sess., June 26, 1922, H. Report 1142, Serial 7957; *Akron* (CO) *Weekly Pioneer Press*, June 2, 1922, 1; Bureau of Land Management, US Department of the Interior, Desert Land Segregations, Patent No. 921824, October 26, 1923.
28. Colorado Secretary of State, Business Center, History and Documents, "Certificate of Incorporation of the Mutual Carey Ditch and Reservoir Company," January 15, 1925, accessed April 24, 2016, www.sos.state.co.us; Laws of Colorado, 1925, Chapter 158; Richard D. O' Brien, interview with the author, September 7, 2007; Isabel O'Brien, interview with the author, September 9, 2001; Carey Act and Desert Land Patents, Segregation No. 11, Patents 1–127, V. 1, Container 12571N, Patents 1–213, V. 2, Container 12571E, SBLC-CSA.
29. Benj. Griffith, *Report of the Colorado Irrigation District Finance Commission to the Twenty-Third General Assembly [of Colorado, 1921]*, 6–7; Laws of Colorado, 1921, Chapters 157–160, Acts; State Engineer of Colorado, *Biennial Report, 1925–1926*, 39–40.
30. Pisani, *Water and American Government*, 136–153; Kluger, *Turning on Water with a Shovel*, 111–113, 115, 118–120, 126; Cannon, "We Are Now Entering a New Era."
31. State Engineer of Colorado, *Biennial Report, 1927–1928*, 40–41; Richard D. O'Brien, interview with the author, September 7, 2007.
32. State Engineer of Colorado, *Biennial Report, 1929–1930*, 42–43; Bent County, Colorado Irrigation District, Resolution, October 30, 1937, Carey Act No. 11 File, Bureau of Land Management, US Department of the Interior, Lakewood, CO.
33. Wickens, *Colorado in the Great Depression*, 176–182; State Engineer of Colorado, *Biennial Report, 1933–1934*.
34. Laws of Colorado, Second Extraordinary Sess., December 4, 1933–January 22, 1934, Chap. 12; *Bent County* (CO) *Democrat*, December 7, 1934, 1; *Commercial and Financial Chronicle* 139 (November–December 1934), 3671.
35. *Bent County* (CO) *Democrat*, December 7, 1934, 1.
36. Bent County, Colorado Irrigation District, Resolution, October 30, 1937, Carey Act No. 11 File, Bureau of Land Management, US Department of the Interior, Lakewood, CO.
37. A. B. Campbell to Commissioner of the General Land Office, October 24, 1934, Carey Act No. 11 File, Bureau of Land Management, US Department of the Interior, Lakewood, CO.
38. Thomas J. Wilson to Secretary of the Interior, June 11, 1938, and Thomas J. Wilson to Franklin D. Roosevelt, June 25, 1938, Carey Act No. 11 File, Bureau of Land

Management, US Department of the Interior, Lakewood, CO; Land Board Minutes Concerning Desert Lands, Container 12571B, List 11, SBLC-CSA.

39. Historical Diversion Records, Water Commissioner's Field Books, Dist. 67, Muddy Creek Reservoir, 1943–1945, OCSE; Carey Act and Desert Land Patents, Segregation No. 11, Patents 1–213, V. 2, Container 12571E, SBLC-CSA. An unsuccessful suit and its unsuccessful appeal to the Colorado Supreme Court, brought decades after a complaint by creditors of the irrigation district, details the final stages of the project's collapse. See In re Bent County, Colorado Irr. Dist. McDermott et al. v. Bent County, Colorado Irr. Dist.

CHAPTER 8: DEFEATED AND BURIED IN DUST

1. Fred L. Harris to Anna B. Loomis, September 25, 1919, TBM.
2. Baca County Historical Society, *Baca County*, 81–84.
3. Baca County Historical Society, *Baca County*, 81–84; Fred L. Harris to Anna B. Loomis, September 25, 1919, TBM; Fred L. Harris, Yearly Precipitation and Reservoir Record Book, TBM; Historical Diversion Records, Water Commissioner's Field Books, Dist. 67, Two Buttes Reservoir, 1915 and 1916, OCSE.
4. List No. 7, State of Colorado for the Two Buttes Irrigation and Reservoir Company, State Selection, Serial No. 01190, Bureau of Land Management, US Department of the Interior, Lakewood, CO; *Colorado Transcript* (Golden), February 24, 1916, 7; Bureau of Land Management, US Department of the Interior, Desert Land Segregations, Patents No. 572557 and No. 66112.
5. Desert Land Certificates and Locations, List No. 7, Container 12571K, SBLC-CSA; Fred L. Harris to John C. Hart, September 22, 1927, TBM; Fred L. Harris to J. B. Marcellus, March 10, 1937, TBM; General Ledger Book, Two Buttes Irrigation and Reservoir Company, TBM; *Moody's Manual of Railroads and Corporation Securities*, 603.
6. *Record Journal of Douglas County* (CO), August 17, 1917, 6; Baca County Historical Society, *Baca County*, 81–82; Harper, "Development of a High Plains Community," 48–50; Colorado State Board of Immigration and State Planning Division, *Year Book of the State of Colorado*, 1920, 190–191, 189, 196; Historical Diversion Records, Water Commissioner's Field Books, Dist. 67, Two Buttes Reservoir, 1919, OCSE.
7. Fred L. Harris to Anna B. Loomis, March 11, 1922, TBM; Obituary of Fred L. Harris, [1940], TBM.
8. Ervin, "Irrigation under the Provisions of the Carey Act," 8–14.
9. Ervin, "Irrigation under the Provisions of the Carey Act," 8–14.
10. US General Land Office, *Annual Report, 1920*, 47–50.
11. Fred L. Harris to Anna B. Loomis, March 24, 1925, TBM; Fred L. Harris to Henry H. Pahlman, December 30, 1927, TBM.
12. Fred L. Harris to Anna B. Loomis, March 24, 1925, TBM; Obituary of Fred L. Harris, [1940], TBM.
13. Fred L. Harris to Bondholders, June 20, 1927, TBM.
14. Fred L Harris to John C. Hart, September 22, 1927, TBM.

15. Fred L. Harris v. Two Buttes Irrigation and Reservoir Company and Others, TBM; Fred L. Harris to Henry H. Pahlman, December 30, 1927, TBM.
16. Fred L. Harris to Henry H. Pahlman, June 20, 1928, TBM; Statement of Fred L. Harris, December 29, 1929, Two Buttes Reservoir and Reservoir Company Foreclosure, TBM.
17. Sheflin, *Legacies of Dust*, 1–4.
18. Fred L Harris to Anna B. Loomis, December 16, 1927, TBM.
19. Harvey, "Creating a 'Sea of Galiea'"; Fred L. Harris to A. B. Loomis, December 26, 1928, TBM; Fred L. Harris to L. S. McMillan, May 24, 1929, TBM.
20. Wickens, *Colorado in the Great Depression*, 16, 22, 30–31, quotation on p. 31.
21. Colorado Office of the Secretary of State, Abstract of Votes Cast at the Primary and General Elections, 1932, comp. (Denver, 1932).
22. Harper, "Development of a High Plains Community," 83–101, 231.
23. Fred L. Harris to Anna B Loomis, December 21, 1929, TBM.
24. Fred L. Harris to L. B. McMillan, February 11, 1930, TMB.
25. Hill, "History of Baca County," 128–130; Colorado State Board of Immigration and State Planning Division, *Year Book of the State of Colorado*, 1930, 67, 69, 76, 81; 1931, 70, 73, 81, 86; 1932, 82; Harper, "Development of a High Plains Community," 90–91; Fred L. Harris to L. B. McMillan, February 11, 1930, TMB; Hewes, *Suitcase Farming Frontier*, 73–81.
26. Fred L. Harris to Maude Harroun, March 10, 1935, TBM.
27. Sheflin, *Legacies of Dust*, 45–71, 73–122.
28. Harper, "Development of a High Plains Community," 123.
29. Cited in Sheflin, *Legacies of Dust*, 73–122; Hurt, "Federal Land Reclamation in the Dust Bowl," 94–106.
30. Harper, "Development of a High Plains Community," 123; Sheflin, *Legacies of Dust*, 91–122, 296–298; Fred L. Harris to E. J. Vonderen, February 24, 1936, TBM; Fred L. Harris to Anna Loomis, November 17, 1937, TBM.
31. Desert Land Certificates of Locations, Container 12571K, List 7, SBLC-CSA; Fred L. Harris to Arthur C. Gordon, April 15, 1937, TBM.
32. *Lamar* (CO) *Daily News*, December 31, 1940, 1, 3.

CHAPTER 9: THE MAKING OF WILDLIFE AREAS

1. White, "Trashing the Trails," 27.
2. *Springfield* (CO) *Herald*, April 9, 1915, 1; October 1, 1915; State Game and Fish Commissioner of Colorado, *Biennial Report, 1913–1914*, 18, 1918–1922, 7–12; Wiltzius, *Fish Culture and Stocking in Colorado*, 91; *Wray* (CO) *Rattler*, June 29, 1933, 6.
3. *Pueblo Indicator*, December 21, 1929, 2; State Game and Fish Commissioner of Colorado, *Biennial Report, 1926–1931*, 26.
4. *Wray* (CO) *Rattler*, April 27, 1933, 5; State Game and Fish Commissioner of Colorado, *Biennial Report, 1939–1941*, 6–7.
5. Fred L. Harris to Arthur C. Gordon, April 15, 1937, and Fred L. Harris to Anna B. Loomis, November 17, 1937, both in TBM; Sheflin, *Legacies of Dust*, 73–122.

6. Laws of Colorado, 1931, An Act, Two Buttes State Game Refuge, 416–418, An Act, Carrizo State Game Refuge, 414–415; In the Matter of the Adjudication of Priorities of the Right to the Use of Water for Domestic and Irrigation Purposes in Water District 67, in the State of Colorado, June 3, 1922, Two Buttes Irrigation and Reservoir System. The Carrizo State Game Refuge continued its designation until 1964, when the Colorado Game and Fish Commission canceled its protection of 7,680 acres after the United States Department of Agriculture, then finalizing the land restoration program, created the Comanche National Grassland. In 2013 the Sikes Ranch State Wildlife Area containing 7,061 acres contiguous to federal land, Colorado State Land Board property, and a Nature Conservancy easement, expanded the spectacular Carrizo habitat. See "Grants to Preserve 15,000 Acres, Create New Public Open Space, Fishing Area," *Great Outdoors Colorado Enewsletter*, June 23, 2014, accessed May 4, 2016, goco.org.
7. *Record Journal of Douglas County* (CO), October 23, 1931, 2. The other game refuges located (but since abandoned) on the plains were Antelope Refuge north of Fort Collins and Smith Hollow Refuge east of Pueblo. See Colorado State Planning Division, *Year Book of the State of Colorado*, 1932 (Denver, 1932), 51–52. Colorado's initial state game refuge law created the Colorado State Game Refuge, located near Rocky Mountain National Park; see Laws of Colorado 1919, Chapter 99.
8. *Record Journal of Douglas County* (CO), March 17, 1933, 4; *Rocky Mountain News*, February 10, 1933, 3.
9. *Record Journal of Douglas County* (CO), March 17, 1933, 4; *Rocky Mountain News*, February 10, 1933, 3; Laws of Colorado, 1933, 521, 524, 531, 534, 538; Maitland v. People.
10. Fred L. Harris to Anna B. Loomis, December 21, 1929; June 6, 1936, both in TBM; Philpott, *Vacationland*, 193–194.
11. Orsi, "From Horicon to Hamburgers," 20–30.
12. Sanderson and Moulton, *Wildlife Issues in a Changing World*, 290–292; Barrows and Holmes, *Colorado's Wildlife Story*, 15–57.
13. Garone, *Fall and Rise of the Wetlands of California's Great Central Valley*, 141–142; Giese, "Federal Foundation for Wildlife Conservation," 117–120, 154–159, 164–168; Fischman, *National Wildlife Refuges*, 32–53. The Pittman-Robertson Act—the most important wildlife funding bill in US history—went into effect on July 1, 1938. Like the Federal Highway Act (1921), it allocated funds raised by the tax to each participating state based on a combination of land area and hunting license sales. For a state to receive eligibility for funds, it had to end all diversions of hunting license revenues to non-wildlife conservation purposes. The federal funds financed state-level wildlife research programs, provided for federal financing of state wildlife refuges and public shooting grounds, and financed other wildlife conservation activities—all of which were subject to supervisory approval of the United States Biological Survey (later the United States Fish and Wildlife Service). In 1941 the United States Congress raised this excise tax to 11 percent. To date, according to the United States Department of the Interior, P-R and its subsequent iterations have allocated more than $22.9 billion to states for wildlife conservation and restoration; state contributions have exceeded $7.6 billion. See Press Release, March 19, 2020 Interior Press, accessed

January 25, 2021, http//www.ios.doi.gov/pressrleases/sportsmen-generate-nearly-1-billion-conservation-funding.

14. Laws of Colorado 1939, An Act, 379; Meine, *Aldo Leopold*, 308–339; Wolf, *Arthur Carhart*, 60–79; Barrows and Holmes, *Colorado's Wildlife Story*, 66–67.
15. *Steamboat Pilot*, November 6, 1947, 7, December 11, 1947, 8, January 1, 1948, 1; *Colorado Conservation Comments* 10, 8 (December 1948), 10.
16. Hooper, *Wetlands of Colorado*, 7–11, 19–21, 35–42; Rutherford, *Canada Geese of Southeastern Colorado*, 1–6, 56–57; State of Colorado, Department of Game and Fish, "Quarterly Progress Report," April 1952, Game Bird Surveys (Project W-37-R-5), 24–33; July 1952, Game Bird Surveys (Project W-37-R-5), 134–141; April 1953, Game Bird Surveys (Project W-37-R-6), 63–70.
17. Warranty Deed, Thomas E. Kitzmiller and Cecile Kitzmiller to the State of Colorado for the benefit of the Game and Fish Commission, October 23, 1957, Colorado Division of Parks and Wildlife, Denver Archive; *Douglas County* (CO) *News*, November 27, 1958, 14; Meyers, "Hot Spot during Dry Spell."
18. Comprehensive fish studies of Two Buttes are held at the Colorado Parks and Wildlife Research Library, Fort Collins. See Lynch, Buscemi, and Lemons, "Limnological and Fishery Conditions of Two Buttes Reservoir"; Lynch, "Fishery Potentials of Irrigation Impoundments"; Lynch, "Progress and Evaluation Report"; Lynch, "Growth Data on Fourteen Fish Species"; Lynch and Taliaferro, "Influence of Irrigation Water Storage and Release Operations."
19. *Report on Sport Fish Restoration Statistical Supplement*, 58–59; *Federal Aid in Fish and Wildlife Restoration*, 13–23; Laws of Colorado 1951, An Act, 431.
20. Colorado Game and Fish Commission, *Colorado Conservation Comments* 10, 8 (December 1948), 11; M. C. Hinderlider, "Report on Muddy Creek Reservoir," September 25, 1950, Muddy Creek Dam File No. 2, Office of Colorado State Engineer, Division of Water Resources, Pueblo Office; *Steamboat* (CO) *Pilot*, October 29, 1953, 10.
21. Richard D. O'Brien, interview with the author, September 7, 2007; Orley O. Phillips to Thomas L. Kimball, June 12, 1957, and Thomas L. Kimball to Ivan C. Crawford, April 14, 1958, both in Muddy Creek (Carey) Dam File No. 3212.51, Office of Colorado State Engineer, Division of Water Resources, Pueblo Office.
22. Orley O. Phillips to Thomas L. Kimball, June 12, 1957, and Thomas Kimball to Ivan C. Crawford, April 14, 1958, both in Muddy Creek (Carey) Dam File No. 3212.51, Office of Colorado State Engineer, Division of Water Resources, Pueblo Office.
23. Thomas Kimball to Ivan C. Crawford, April 14, 1958, Muddy Creek (Carey) Dam File No. 3212.51, Office of Colorado State Engineer, Division of Water Resources, Pueblo Office; Bent County District Court, State of Colorado, In the Matter of the Adjudication of Priorities, May 29, 1959; Las Animas County District Court, State of Colorado, In the Matter of the Adjudication of Priorities, May 20, 1963.
24. Moser, "Lake Setchfield."
25. Matthai, *Floods of June 1965 in South Platte River Basin*; Snipes and others, *Floods of June 1965 in the Arkansas River Basin*, 1, 27, 29; *Plainsman Herald* (Springfield, CO), June 24, 1965, 1; December 9, 1965, 1; *Denver Post*, June 19, 1965, 3; *Bent County* (CO) *Democrat*, June 24, 1965, 1; July 8, 1965, 1.

26. Harold E. Eyrich to E. L. Shaw, November 4, 1965, Muddy Creek (Carey) Dam File No. 3212.51, Office of Colorado State Engineer, Division of Water Resources, Pueblo Office.
27. Hoffman, *Scaled Quail in Colorado*, 6–46.
28. George E. Kimble, Colorado Legislator Record, accessed March 10, 2014, www.leg.state.co.us/lcs/leghist.nsf; *Douglas County* (CO) *News*, April 21, 1955; *Colorado Outdoors* 20, 1 (January–February 1971), 44; 22, 2 (March–April 1973), 23; 22, 3 (May–June 1973), 44–45.
29. *Colorado Outdoors* 22, 3 (May–June 1973), 44–45; Welch, *Mission in the Desert*, 72–75.
30. Kimble, "Comments," *Outdoor America* 24, 11 (November 1959), 2–3; Welch, *Mission in the Desert*, 72–75.
31. US Public Law 89:298; Resolution Concerning John Martin Reservoir Permanent Pool, accessed June 11, 2015, water.state.co.us/DWRIPub/Documents/1976JMRPermPoolResolution.pdf; In the Matter of the Petition of the Colorado Game, Fish, and Parks, June 13, 1968.
32. Resolution Concerning John Martin Reservoir Permanent Pool, August 14, 1976, accessed June 11, 2015, water.state.co.us/DWRIPub/Documents/1976JMRPermPoolResolution.pdf; In the Matter of the Application for Change of Water Rights of the State of Colorado, November 2, 1979.
33. Paul H. Berg to State Engineer of Colorado, September 17, 1965, Two Buttes File, Office of Colorado State Engineer, Division of Water Resources, Pueblo Office; Warranty Deed, Two Buttes Mutual Water Association to the State of Colorado through the Department of Natural Resources, August 13, 1970, Colorado Division of Wildlife, Denver Archive; *Plainsman Herald* (Springfield, CO), August 27, 1970, 1.

CHAPTER 10: SUSTAINING A CONSERVATION PURPOSE

1. Warren, "Collecting Trip to Southeastern Colorado."
2. Armstrong, "Edward Royal Warren (1862–1942)."
3. Warren, *Mammals of Colorado*.
4. Quoted in Armstrong, "Edward Royal Warren (1862–1942)," 367.
5. Schulte, *Wayne Aspinall and the Shaping of the American West*, 115–176; Cronon, "The Trouble with Wilderness."
6. For a list and map of all LWCF project grants in Colorado, see Colorado Parks and Wildlife, Land and Water Conservation Fund, accessed March 4, 2022, cpw.state.co.us.
7. "Environmental Assessment Report," Project: W-109-D, Colorado Division of Wildlife, Southeast Properties [1979–1980], Colorado Parks and Wildlife Research Library, Fort Collins.
8. Southeast Region Property Development Project: W-109-D.
9. Southeast Region Property Development Project: W-109-D.
10. Colorado Parks and Wildlife, "Light Geese FAQs," accessed September 14, 2022, cpw.state.co.us; Peterson, Pieplow, and Spencer, "Birding in Colorado."

11. Sheflin, *Legacies of Dust*, 73–122.
12. Barrows and Holmes, "Colorado's Wildlife Story," 260; Warren, *Mammals of Colorado*; "Herd Counts."
13. Phase I Inspection Report, National Dam Safety Program, "Two Buttes Dam Baca County, Colorado"; "Two Buttes Lake Dam Hydrologic Study, Baca County Colorado."
14. *Lamar* (CO) *Ledger*, September 19, 2012, 1; *Kiowa County* (CO) *Press*, August 25, 2018, 1.
15. Schulte, *As Precious as Blood*, 65–72, 81–119.
16. Schulte, *As Precious as Blood*, 255–261; Bassi, Schneider, and White, "ISF Law."
17. Young, Olmstead, and Boldt, "Environmental Assessment Report."
18. Young, Olmstead, and Boldt, "Environmental Assessment Report."
19. Young, Olmstead, and Boldt, "Environmental Assessment Report."
20. "Herd Counts."
21. *Little Snake Resource Management Plan and Final Environmental Impact Statement*; Elliott and Gyetvai, "Channel-Pattern Adjustments and Geomorphic Characteristics of Elkhead Creek, Colorado."
22. Playa Lakes Joint Venture, at www.pljv.org; Colorado Natural Heritage Program, at cnhp.colostate.edu.
23. Daltry, Bonada, and Boulton, *Intermittent Rivers and Ephemeral Streams*.
24. Southeastern Colorado Water Conservancy District, at www.secwcd.org; Purgatoire River Water Conservancy District, at www.prwcd.com; Lower Arkansas River Conservancy District, at www.lavwcd.com.

BIBLIOGRAPHY

ARCHIVE AND MANUSCRIPT COLLECTIONS

Alaska State Library. Frank H. Whiting Photograph Collection, ca. 1898–1902, PCA 219, Alaska State Library–Historical Collections Finding Aids. Accessed January 25, 2015. https://alaska.libraryhost.com/repositories/2/resources/126.

City of Greeley Museum, Greeley, CO. Permanent Collections.

Colorado Parks and Wildlife Archive, Denver, CO.

Colorado Parks and Wildlife Research Library, Fort Collins, CO. Two Buttes Collection.

Colorado State Archives, Denver. Records of the State Board of Land Commissioners of Colorado, Carey Act Records.

Reports on Carey Act Selection No. 11, Bent County, Colorado, February 17, 1922, Box 10256.

Dr. Verity Museum, Two Buttes, CO. Bent and Prowers Irrigation District Report. Lamar, 1912. Two Buttes Reservoir and Irrigation Company Records.

Gerald C. Morton Collection, Fort Collins, CO. Isabel O'Brien, "Spouse and Me." O'Brien Family Photographs, author's collection.

History Colorado, Denver. Arkansas Valley Ditch Association, MSS#16.

History Colorado, Denver. Colorado Millennial Site/Hackberry Site/Bloody Springs, National and State Register, April 8, 1980, Site 5LA.1115.

https://doi.org/10.5876/9781646426492.c013

Idaho State Historical Society, Boise. Twin Falls Land and Water Company Files, MS114 General Correspondence (Secretary's Files).

Museum of Northwest Colorado, Craig. Wayne Wymore Collection, 04–27.

National Archives, Denver Branch, Denver, CO. Letters Received, Records of the GLO, Record Group 49.

National Archives, Washington, DC. Carey Act Cases, Colorado-Idaho, GLO, Record Group 49.

Office of Colorado State Engineer (Division of Water Resources), Denver Office, Denver, CO. Map and Filing Statements, Water Commissioner's Reports and Field Books and Water Rights.

Office of Colorado State Engineer (Division of Water Resources), Pueblo District Office, Pueblo. Muddy Creek Dam and Two Buttes Dam Files.

Phase I Inspection Report, National Dam Safety Program. "Two Buttes Dam Baca County, Colorado." US Corps of Engineers ID #CO 759, August 1980.

"Two Buttes Lake Dam Hydrologic Study, Baca County, Colorado." Colorado Division of Wildlife, Pueblo, November 2009.

US Bureau of Land Management, Archives, Lakewood, CO. Land Records and Serials. Carey Act No. 11 File. Frank H. Whiting Maps.

Water Resources Center Archive, Colorado State University, Fort Collins. Papers of Delph E. Carpenter.

Western History Collection, Denver Public Library, Denver, CO.

Donovan, John Philip. "Report on Badito Irrigation Project." Donovan Papers, CMSS-M27.

Perrine, Frederick Auten Combs. "Report on Power Development of the Huerfano River at Badito, Huerfano County, Colorado, 1907," C MSS-M28.

Wyoming State Archives, Cheyenne. Photographs.

INTERVIEWS WITH THE AUTHOR

O'Brien, Gerald G., June 17, 1976, June 11, 1982, La Junta, CO.

O'Brien, Isabel Dodge, September 9, 2001, Denver, CO.

O'Brien, Richard D., September 7, 2007, Aurora, CO.

PUBLIC DOCUMENTS OF THE UNITED STATES

Agricultural Statistics: Crops and Livestock of the State of Colorado, 1931. Bulletin 89, United States Department of Agriculture, Bureau of Agricultural Economics in Cooperation with the Colorado State Board of Immigration, 1932.

Beach, C. W., and P. J. Preston. *Irrigation in Colorado*. US Department of Agriculture Office of Experiment Stations, Bulletin 218. Washington, DC: Government Printing Office, 1910.

Carey Act Projects. 62d Cong., 3d Sess., February 21, 1913, S. Doc. 1097 (Serial 6365).

Colorado Enabling Act. 43d Cong., 2d Sess., March 3, 1875, Ch. 139.

Congressional Record.

Department of Agriculture, Annual Reports of the Office of Experiment Stations.

Department of Commerce, Bureau of Census.

Department of the Interior, Annual Reports.

Ervin, Guy. "Irrigation under the Provisions of the Carey Act." US Department of Agriculture Circular 124. Washington, DC: Government Printing Office, 1919.

Gates, Paul W. *History of Public Land Law Development.* Washington, DC: Government Printing Office, 1968.

General Land Office, Annual Reports.

Geological Survey, Annual Reports.

Grant of Certain Lands to Colorado. 59th Cong., 2d Sess., February 25, 1907, S. Report 7279 (Serial 5061).

Hinton, Richard J. "Irrigation in the United States." 49th Cong., 2d Sess., 1887, S. Misc. Doc. 15 (Serial 2450).

Hutchins, Wells A. "Irrigation Districts, Their Operation and Financing." US Department of Agriculture Technical Bulletin 254 (June 1934): 1–93.

Hutchins, Wells A. "Mutual Irrigation Companies." US Department of Agriculture Technical Bulletin 82 (January 1929): 1–50.

Little Snake Resource Management Plan and Final Environmental Impact Statement, for Public Lands Administered by the Bureau of Land Management Little Snake Field Office, Craig, Colorado. 3 vols. Washington, DC: US Department of the Interior, August 2010.

Matthai, H. F. *Floods of June 1965 in South Platte River Basin, Colorado.* US Geological Survey Water Supply Paper 1850-B. Washington, DC: Government Printing Office, 1969.

McLaughlin, Thad B. *Geology and Ground Water Resources of Baca County, Colorado.* US Geological Survey Water Supply Paper 1256. Washington, DC: Government Printing Office, 1954.

Newell, Frederick H. *Report on Agriculture by Irrigation in the Western Part of the United States at the Eleventh Census: 1890.* Washington, DC: Government Printing Office, 1894.

Patent of Carey Act Segregation List *No. 11 to the State of Colorado.* 67th Cong., 2d Sess., June 26, 1922, H. Report 1142 (Serial 7957).

Preference Right of Entry by Certain Carey Act Entrymen. 66th Cong., 1st Sess., August 2, 1919, H. Doc. 219 (Serial 7592).

Preference Right of Entry by Certain Carey Act Entrymen. 66th Cong., 2d Sess., January 8, 1920, S. Report 357 (Serial 7649).

Press Release. Interior Press, March 19, 2020. Accessed January 25, 2021. http//www.ios.doi.gov/pressrleases/sportsmen-generate-nearly-1-billion-conservation-funding.

Public Laws.

Reclamation Service, Annual Reports.

Relief to Settlers under Forfeited Carey Act Projects. 65th Cong., 2d Sess., January 25, 1918, H. Doc. 259 (Serial 7307).

Relief to Settlers under Forfeited Carey Act Projects. 65th Cong., 3d Sess., February 6, 1919, S. Doc. 686 (Serial 7452).

Report of the Special Committee of the United States Senate on the Irrigation of Arid Lands, vol. 3, Rocky Mountain Region and Great Plains. 51st Cong., 1st Sess., 1890, S. Report 928, Part 4 (Serial 2708).

Scott, Glenn R. *Geologic and Structure Contour Map of the La Junta Quadrangle, Colorado and Kansas*. Washington, DC: US Geological Survey, 1968.

Segregated Lands under the Public Domain. 62d Cong., 3d Sess., 1913, S. Report 115 (Serial 6360).

Snipes, R. J., and others. *Floods of June 1965 in the Arkansas River Basin, Colorado, Kansas, and New Mexico*. US Geological Survey Water Supply Paper 1850-D. Washington, DC: Government Printing Office, 1974.

Statutes at Large.

Stover, A. P. "Irrigation under the Carey Act." In US Department of Agriculture, *Annual Report of the Office of Experiment Stations for the Year Ending June 30, 1910*, 461–488. Washington, DC: Government Printing Office, 1911.

US Department of the Interior. *Bureau of Reclamation Project Feasibilities and Authorizations: A Compilation of Findings of Feasibilities and Authorizations for Bureau of Reclamation Projects of the Department of Interior*. Washington, DC: Government Printing Office, 1957.

Wieber, F. W. F. "The History of the United States Naval Hospital, Fort Lyon, Colo., and the Activities of the Medical Corps in the Development of the Hospital for Sanatorium Purposes." *United States Naval Medical Bulletin* 17, no. 5 (November 1922): 745–757.

Wildlife Restoration and Conservation, Proceedings of the North American Wildlife Conference Called by President Franklin D. Roosevelt, Printed for the Use of the Special Committee on Conservation of Wildlife Resources. 74th Cong., 2d Sess., Committee Print. Washington, DC: Government Printing Office, 1936.

PUBLIC DOCUMENTS OF THE STATE OF COLORADO

Bray, Charles I. *Financing Western Cattlemen*. Colorado Agricultural College, Colorado Experiment Station, Bulletin 338. Fort Collins: Colorado Agricultural College, 1928.

Colorado Constitution.

Colorado Game and Fish Commission, *Conservation Comments*.

Department of Game and Fish, Quarterly Progress Reports (Pittman-Robertson, Dingell-Johnson Projects).

Griffith, Benj. *Report of the Colorado Irrigation District Finance Commission to the Twenty-Third General Assembly [of Colorado, 1921]*.

Hoffman, Donald M. *The Scaled Quail in Colorado: Range, Population Status, Harvest*. Denver: State of Colorado, Division of Game, Fish and Parks, June 1965.

Hooper, Richard. *Wetlands of Colorado*. Denver: State of Colorado, Division of Game, Fish and Parks, September 1968.

Laws of Colorado.

Lynch, Thomas M. "The Fishery Potentials of Irrigation Impoundments in the Lower Arkansas River Drainage Area of Colorado." Special Purpose Report 20. Colorado Game and Fish Commission, January 17, 1955. Colorado Parks and Wildlife Research Library, Fort Collins, CO.

Lynch, Thomas M. "Growth Data on Fourteen Fish Species Collected from the Warm Water Regions of Colorado." Special Purpose Report 48. Colorado Game and Fish Commission, April 1, 1957. Colorado Parks and Wildlife Research Library, Fort Collins, CO.

Lynch, Thomas M. "A Progress and Evaluation Report on the Success of the Walleye (*Stizostedium vitreum*) in Colorado." Special Purpose Report 31. Colorado Game and Fish Commission, June 8, 1955. Colorado Parks and Wildlife Research Library, Fort Collins, CO.

Lynch, Thomas M., Philip A. Buscemi, and David G. Lemons. "Limnological and Fishery Conditions of Two Buttes Reservoir, Colorado 1950 and 1951." No report number. State Game and Fish Commission, November 10, 1953. Colorado Parks and Wildlife Research Library, Fort Collins, CO.

Lynch, Thomas M., and Rex I. Taliaferro. "The Influence of Irrigation Water Storage and Release Operations on Warm Water Sport Fishing in Two Colorado Impoundments." Report II. Colorado Game and Fish Commission, [1960]. Colorado Parks and Wildlife Research Library, Fort Collins, CO.

McMahon Audit Company. "Financial Statement of the Mutual Carey Irrigation Company, January 31, 1922." Box 10256: Reports on Carey Act Selection No. 11, Bent County, Colorado, February 17, 1922, 26–33. Colorado State Archives, Records of the State Board of Land Commissioners of Colorado.

Murphy, Edward Charles, and others. *Destructive Floods in the United States in 1904*. US Geological Survey Water Supply Paper 147. Washington, DC: Government Printing Office, 1905.

Murphy, Will R. "A History of Carey Act Selection No. 11, Bent County, Colorado and Suggestions for Completion of Financing, February 17, 1922." Box 10256: Reports on Carey Act Selection No. 11, Bent County, Colorado, February 17, 1922, 7–12. Colorado State Archives, Records of the State Board of Land Commissioners of Colorado.

Olmstead, Bill. "Environmental Assessment Report, Project: W-108-D." Southeast Properties [1979–1980]. Colorado Parks and Wildlife Research Library, Fort Collins, CO.

Rutherford, William H. *The Canada Geese of Southeastern Colorado*. Denver: State of Colorado, Division of Game, Fish and Parks, September 1970.

Schrontz, C. C. "Report of the Cost of Completed Construction and Costs of Completion." Box 10256: Reports on Carey Act Selection No. 11, Bent County, Colorado, February 17, 1922, 13–25. Colorado State Archives, Records of the State Board of Land Commissioners of Colorado.

Secretary of State, Abstract of Votes Cast at the Primary and General Elections.

Secretary of State, *Biennial Reports*.

Secretary of State, Business Center, History and Documents.

Southeast Region Property Development Project: W-109-D. Colorado Division of Wildlife, Period: June 1, 1979, to May 31, 1982. Colorado Parks and Wildlife Research Library, Fort Collins, CO.

State Board of Immigration and State Planning Division, *Year Books*.

State Board of Land Commissioners of Colorado, *Biennial Reports*.

State Board of Land Commissioners of Colorado, Rules and Regulations of the State Board of Land Commissioners of Colorado in Relation to the Entry of Land under the Provisions of the Carey Act (Revised to March 1, 1919).

State Engineer, *Biennial Reports.*
State Game and Fish Commissioner, *Biennial Reports.*
Wiltzius, William J. *Fish Culture and Stocking in Colorado, 1877–1978.* Colorado Division of Wildlife Report 12, June 1985. Denver: Colorado Division of Wildlife, 1985.
Young, James, William Olmstead, and Wilbur Boldt. "Environmental Assessment Report on the Proposed Acquisition of 940 Acres of Land Plus a 10 Year Recreation Easement on 6,000 Acres in Exchange for 2,438.08 Acres of Division of Wildlife Property Known as Setchfield Dam." Colorado Division of Wildlife, Lamar Area Office, January 1982.

JOINT PUBLIC DOCUMENTS COLORADO–UNITED STATES

Elliott, John G., and Stevan Gyetvai. "Channel-Pattern Adjustments and Geomorphic Characteristics of Elkhead Creek, Colorado, 1937–97." Water Resources Investigations Report 99-4098. US Department of the Interior, US Geological Survey, prepared in cooperation with the Colorado River Water Conservation District, Denver, 1999.

COURT PROCEEDINGS OF THE UNITED STATES

Greer v. Connecticut, 161 U.S. 519 (1896).
Kansas v. Colorado, 206 US 46 (1907).

COURT PROCEEDINGS OF THE STATE OF COLORADO

Fred L. Harris v. Two Buttes Irrigation and Reservoir Company and others, Baca County District Court, Colo. (August 12, 1927).
In re Bent County, Colorado Irr. Dist. McDermott et al. v. Bent County, Colorado Irr. Dist., Case No. 17214, 130 Colo. 44, 272 P. 2d 995 (July 6, 1954).
In re Salaries of Commissioners and Employees of State Land Board, 55 Colo. 105 (June 27, 1913).
In the Matter of the Adjudication of Priorities of the Right to the Use of Water for Domestic and Irrigation Purposes in Water District 19, in the State of Colorado, January 12, 1925, Smith Canon Canal/Reservoir, Case No. CA6118, Las Animas County District Court (January 12, 1925), 346–348, 582–585.
In the Matter of the Adjudication of Priorities of the Right to the Use of Water for Domestic and Irrigation Purposes in Water District 67, in the State of Colorado, June 3, 1922, Muddy Creek Canal/Reservoir, Case No. 06/03/1922, Bent County District Court (June 3, 1922), 445 1/2-448, 494–497, 505–508, 518–523.
In the Matter of the Adjudication of Priorities of the Right to the Use of Water for Domestic and Irrigation Purposes in Water District 67, in the State of Colorado, June 3, 1922, Two Buttes Irrigation and Reservoir System, Case No. 06/03/1922, Bent County District Court (June 3, 1922), 491 1/2-493.
In the Matter of the Adjudication of Priorities of the Right to the Use of Water for Domestic and Irrigation Purposes in Water District No. 67, in the State of Colorado, May 29,

1959, Muddy Creek Canal/Reservoir, Case No. CA0418, Bent County District Court (May 29, 1959), 935–939.
In the Matter of the Adjudication of Priorities of the Right to the Use of Water for Domestic and Irrigation Purposes in Water District 19, in the State of Colorado, May 20, 1963, Smith Canon Canal/Smith Canon Reservoir, Case No. CA6118–63, Las Animas County District Court (May 20, 1963), 0890–0894.
In the Matter of the Application for Change of Water Rights of the State of Colorado, Department of Natural Resources, Division of Wildlife, in Muddy Creek, a Tributary of Rule Creek and the Arkansas River in Bent County, November 2, 1979, Findings of Fact, Conclusions of Law, and Decree, Case No. W-4605, District Court in and for Water Division 2 (November 2, 1979).
In the Matter of the Petition of the Colorado Game, Fish, and Parks Commission to Change the Point of Location for Storing Water of the Muddy Creek Reservoir Appropriation Priority No. 46 1/2, Reservoir Priority No. 8, in Water District No. 67, Irrigation Division No. 2, in the State of Colorado, June 13, 1968, Findings of Fact, Conclusions of Law, and Decree, Civil Action No. 1434, District Court in and for the County of Bent, State of Colorado (June 13, 1968) CRS 37-92-103 (5).
Maitland v. People, 93 Colo. 59 23P, 2d 116 (June 5, 1933).
People ex rel. Murphy v. Field, 66 Colo. 367, 181 (1919).
R. B. McDermott v. the Bent County, Colorado Irrigation District, Case No. 17945, decided March 11, 1957 (135 Colo. 70, 308 P.2d 603; 1957).
Riverside Reservoir and Land Company v. Bijou Irrigation District et al., 65 Colo. 184 (1918).
Routt County Development Company v. Johnston et al., 23 Colo. App. 511 (March 10, 1913).

NEWSPAPERS, NEWSLETTERS, AND MAGAZINES

Akron (CO) *Weekly Pioneer Press*
Aspen Daily Times
Aspen Democrat-Times
Bayfield (CO) *Blade*
Bent County (CO) *Democrat*
Boulder Camera
Brandon (CO) *Bell*
Breckenridge (CO) *Bulletin*
Brick (Chicago trade magazine)
Castle Rock (CO) *Journal*
Chafee County (CO) *Republican*
Cheyenne (WY) *Daily Sun*
Colorado Transcript (Golden)
Commercial and Financial Chronicle
Craig (CO) *Daily Press*
Creede (CO) *Candle*

Daily Boomerang (Laramie, WY)
Denver Farm and Field
Denver Post
Douglas County (CO) *News*
Durango (CO) *Democrat*
Durango (CO) *Wage Earner*
Eagle County (CO) *Blade*
Eagle Valley (CO) *Enterprise*
Electrical World
Fairplay (CO) *Flume*
Forestry and Irrigation
Fort Collins Courier (Daily and Weekly)
Golden (CO) *Transcript*
Great West
Greeley (CO) *Tribune*
Haswell (CO) *Herald*
Ignacio (CO) *Chieftain* (Weekly)
Irrigation Age
Kiowa County (CO) *Press*
Lamar (CO) *Daily News*
Lamar (CO) *Ledger*
Las Animas (CO) *Leader*
Leadville (CO) *Daily and Evening Chronicle*
Montezuma (CO) *Journal*
Moody's Magazine
New York Times
Oak Creek (CO) *Times*
Oregonian
Outdoor America
Plainsman Herald (Springfield, CO)
Pueblo Indicator
Range Leader (Hugo, CO)
Record Journal of Douglas County (CO)
Rocky Mountain News (Denver)
Routt County (CO) *Republican*
Routt County (CO) *Sentinel*
San Juan (CO) *Prospector*
Science
Silver Cliff (CO) *Rustler*
Silverton (CO) *Standard*
Springfield (CO) *Herald*
Steamboat Pilot
Sugar Beet Gazette

Summit County (CO) *Journal*
Telluride Journal
Two Buttes (CO) *Sentinel*
Weekly Ignacio (CO) *Chieftain*
Wray (CO) *Rattler*
Wyoming Weekly Tribune (Cheyenne)
Yampa (CO) *Leader*

INTERNET SOURCES

Birding in Colorado, Birds in Baca County, Colorado. coloradobirding.org.
Colorado Agriculture Bibliography. http://lib.colostate.edu/research/agbib.
Colorado Business Directory, Las Animas, 1911. usgwarchives.net/co/bent/history.
Colorado General Assembly, Legislator Record. www.leg.state.co.us./lcs/leghist.nsf.
Colorado Natural Heritage Program. cnhp.colostate.edu.
Colorado Parks and Wildlife. cpw.state.co.us.
Colorado Secretary of State, Business Center, History and Documents. www.sos.state.co.us.
Colorado Water Conservation Board. cwcb.colorado.gov.
Great Outdoors Colorado Enewsletter. www.goco.org.
Lower Arkansas River Conservancy District. www.lavwcd.com.
Playa Lakes Joint Venture. www.pljv.org.
Purgatoire River Water Conservancy District. www.prwcd.com.
Register of Funerals, 1896–1910. www.cogenweb.com/bent/funeral.
Southeastern Colorado Water Conservancy District. www.secwcd.org.
United States Fish and Wildlife Service. www.fws.gov.

BOOKS, ARTICLES, PROCEEDINGS, REPORTS, THESES, AND DISSERTATIONS

Abbott, Carl. *Colorado: A History of the Centennial State*. Boulder: Colorado Associated University Press, 1976.

Abbott, Carl, Stephen J. Leonard, and Thomas J. Noel. *Colorado: A History of the Centennial State*, 4th ed. Boulder: University Press of Colorado, 2005.

Archer, Kenna Lang. *Unruly Waters: A Social and Environmental History of the Brazos River*. Albuquerque: University of New Mexico Press, 2015.

Armstrong, David M. "Edward Royal Warren (1862–1942) and the Development of Coloradan Mammalogy." *American Zoologist* 26, no. 2 (1986): 363–370.

Baca County Historical Society. *Baca County*. Lubbock, TX: Specialty Publishing, 1983.

Barrows, Pete, and Judith Holmes. *Colorado's Wildlife Story*. Denver: Colorado Division of Wildlife, 1990.

Bassi, Linda J., Susan J. Schneider, and Kaylea M. White. "ISF Law—Stories about the Origin and Evolution of Colorado's Instream Flow Law in This Prior Appropriation State."

University of Denver Water Law Review 22, no. 2, 389–436. cwcb.colorado.gov, Instream Flow Program. Accessed October 23, 2022.

Bogener, Stephen. *Ditches across the Desert: Irrigation in the Lower Pecos Valley*. Lubbock: Texas Tech University Press, 2003.

Bonner, Robert E. "Elwood Mead, Buffalo Bill Cody, and the Carey Act in Wyoming." *Montana: The Magazine of Western History* 55, no. 1 (Spring 2005): 36–51.

Bonner, Robert E. *William F. Cody's Wyoming Empire: The Buffalo Bill Nobody Knows*. Norman: University of Oklahoma Press, 2007.

Brokaw, Howard P. *Wildlife and America: Contributions to an Understanding of American Wildlife and Its Conservation*. Washington, DC: Council on Environmental Quality, 1978.

Brookings Institution for Government Research. *The U.S. Reclamation Service: Its History, Activities, and Organization*. New York: D. Appleton, 1919.

Brown, Jen Corrine. *Trout Culture: How Fly Fishing Forever Changed the Rocky Mountain West*. Seattle: University of Washington Press, 2015.

Burroughs, John Rolfe. *Where the Old West Stayed Young*. New York: William Morrow, 1962.

Cannon, Brian Q. "We Are Now Entering a New Era: Federal Reclamation and the Fact Finding Commission of 1923–1924." *Pacific Historical Review* 66, no. 2 (May 1977): 185–211.

Carrillo, Richard F. "An In-Depth Review of Regional History: Summary of the Culture and History of the Purgatoire and Arkansas Valley Region in Southeastern Colorado." 2008. Accessed October 23, 2016. www.secoloradoheritage.com.

Cech, Thomas V. *Principles of Water Resources: History, Development, Management, and Policy*, 3rd ed. Hoboken, NJ: John Wiley and Son, 2009.

Chamberlain, Lawrence. *The Principals of Bond Investment*. New York: Henry Holt, 1911.

Church, Minette C. "Homesteads on the Purgatorie: Frontiers of Culture Contact in Nineteenth Century Colorado." PhD dissertation, University of Pennsylvania, Philadelphia, 2002.

Church, Minette C. "Purgatorio, Purgatoire, or Picketwire: Negotiating Local, National, and Transnational Identities along the Puragtoire River in Nineteenth-Century Colorado." In *Archaeological Landscapes on the High Plains*, ed. Laura L. Scheiber and Bonnie J. Clark, 58–73. Boulder: University Press of Colorado, 2008.

Clark, Bonnie, J. *On the Edge of Purgatory: An Archaeology of Place in Hispanic Colorado*. Lincoln: University of Nebraska Press, 2011.

Clark, Ira G. *Water in New Mexico: A History of Its Management and Use*. Albuquerque: University of New Mexico Press, 1987.

Clason, George S. *Free Homestead Land of Colorado Described: A Handbook*. Denver: Clason, 1915.

Colorado Carey Act Land Opening. Lamar, CO: Two Buttes Irrigation and Reservoir Company, 1909.

Colorado Preservation, Inc. *Cultural Resources Survey of the Purgatoire River Region*. Denver: Colorado Preservation, Inc., 2011. Accessed January 3, 2018. http://coloradopreservation.org/programs.

Cronon, William. *Changes in the Land: Indians, Colonists, and the Ecology of New England*. New York: Hill and Wang, 1983.

Cronon, William. "The Trouble with Wilderness; or, Getting Back to the Wrong Nature."

In *Uncommon Ground: Rethinking the Human Place in Nature*, ed. William Cronon, 69–90. New York: W. W. Norton, 1995.

Daltry, Thibault, Nuria Bonada, and Andrew J. Boulton. *Intermittent Rivers and Ephemeral Streams: Ecology and Management*. Cambridge: Academic Press, 2017.

Davison, Graeme. "Country Life." In *Struggle Country: The Rural Ideal in Twentieth Century Australia*, ed. Graeme Davison and Marc Brodie, 2–4. Randwick, Australia: Monash University Publishing, 2005.

Deutsch, Sarah. *Making a Modern West: The Contested Terrain of a Region and Its Borders, 1848–1940*. Lincoln: University of Nebraska Press, 2022.

Deutsch, Sarah. *No Separate Refuge: Culture, Class, and Gender on an Anglo-Hispanic Frontier*. New York: Oxford University Press, 1989.

Dille, J. M. *Irrigation in Morgan County*. Fort Morgan, CO: Farmers State Bank, 1960.

Directory of Colorado Springs. Colorado Springs: Gazette Publishing, 1896.

Dunbar, Robert G. *Forging New Rights in Western Waters*. Lincoln: University of Nebraska Press, 1983.

Dunlap, Thomas R. *Saving America's Wildlife: Ecology and the American Mind, 1850–1990*. Princeton, NJ: Princeton University Press, 1991.

Federal Aid in Fish and Wildlife Restoration: Annual Report on Dingell-Johnson and Pittman-Robertson Programs of the Fiscal Year Ending June 30, 1974. Washington, DC: Wildlife Management Institute and Sport Fishing Institute, 1975.

"Federal Officers and Western Governors Attending the Irrigation Conference." *Municipal Facts: The City of Denver* 2, no. 15 (May 9, 1914): 8–9.

Fiege, Mark. *Irrigated Eden: The Watering of an Agricultural Landscape*. Weyerhaeuser Environmental Books. Seattle: University of Washington Press, 1999.

Fine-Dare, Kathleen S. *Grave Injustice: The American Repatriation Movement and NAGPRA*. Lincoln: University of Nebraska Press, 2002.

Fishchman, Robert L. *The National Wildlife Refuges: Coordinating a Conservation System through Law*. Washington, DC: Island Press, 2003.

Fite, Gilbert C. *George N. Peek and the Fight for Farm Parity*. Norman: University of Oklahoma Press, 1954.

Flores, Dan. *American Serengeti: The Last Big Animals of the Great Plains*. Lawrence: University Press of Kansas, 2017.

Fort Lyon Canal Company. "145 Years of Rainfall Records for Las Animas, Colorado." Accessed March 15, 2014. www.flcc.net/Rainfall.

Freeman, John F. *High Plains Horticulture: A History*. Boulder: University Press of Colorado, 2008.

Freeman, John F. *Persistent Progressives: The Rocky Mountain Farmers Union*. Boulder: University Press of Colorado, 2016.

Freeman, W. B., W. A. Lamb, and R. H. Bolster. *Surface Water Supply of the United States, 1907–1908*, part 7: *Lower Mississippi Basin*. US Geological Survey Water Supply Paper 247. Washington, DC: Government Printing Office, 1910.

Friedman, Paul D. *Valley of Lost Souls: A History of the Pinon Canyon Region of Southeastern Colorado*. Essays and Monographs in Colorado History 3, 1988. Denver: State Historical Society of Colorado, 1989.

Garone, Philip. *The Rise and Fall of the Wetlands of California's Central Valley*. Berkley: University of California Press, 2011.

Giese, Michael W. "A Federal Foundation for Wildlife Conservation: The Evolution of the National Wildlife System, 1920–1969." PhD dissertation, American University, Washington, DC, 2008.

Giesecke, Kay, Francis A. Longstaff, Stephen Schaefer, and Ilya Stebulaev. "Corporate Bond Default Risk: A 150-Year Perspective." *Journal of Financial Economics* 102, no. 2 (November 2011): 233–250.

Goldberg, Alan. *Hooded Empire: The Ku Klux Klan in Colorado*. Urbana: University of Illinois Press, 1981.

Graham, Frank, Jr. *Man's Dominion: The Story of Conservation in America*. New York: M. Evans, 1971.

Grinnell, George Bird. *The Fighting Cheyennes*. New York: Charles Scribner's Sons, 1915.

Hämäläinen, Pekka. *The Comanche Empire*. New Haven, CT: Yale University Press, 2008.

Hamblin, Jacob Darwin, ed. "Philip Garone, *The Fall and Rise of the Wetlands of California's Great Central Valley* (California, 2011)." *H-Environment Roundtable Review* 4, no. 10 (2014): 1–31.

Hansen, James E., II. *Democracy's College in the Centennial State: A History of Colorado State University*. Fort Collins: Colorado State University, 1977.

Hargreaves, Mary Wilma M. *Dry Farming in the Northern Great Plains: Years of Readjustment, 1900–1925*. Cambridge: Harvard University Press, 1957.

Hargreaves, Mary Wilma M. "Hardy Webster Campbell (1850–1937)." *Agricultural History* 32, no. 1 (January 1958): 62–65.

Harper, Thomas Alan. "The Development of a High Plains Community: A History of Baca County, Colorado." MA thesis, University of Denver, Denver, Colorado, 1967.

Harris, Katherine. *Long Vistas: Women and Families on Colorado Homesteads*. Niwot: University Press of Colorado, 1993.

Harvey, Douglas S. "Creating a 'Sea of Galilea': The Rescue of Cheyenne Bottoms Wildlife Area, 1927–1930." *Kansas History* 24, no. 1 (Spring 2001): 2–17.

Harvey, Mark W. T. *A Symbol of Wilderness: Echo Park and the American Conservation Movement*. Albuquerque: University of New Mexico Press, 1994.

Hays, Samuel P. *Conservation and the Gospel of Efficiency: The Progressive Conservation Movement, 1890–1920*. Cambridge: Harvard University Press, 1959.

Healy, David. "Admiral William B. Caperton: Proconsul and Diplomat." In *Behind the Throne: Servants of Power to Imperial Presidents, 1898–1968*, ed. Thomas J. McCormick and Walter LaFeber, 67–100. Madison: University of Wisconsin Press, 1993.

"Herd Counts." *Colorado Outdoors* (January–February 2023): 33–35.

Hewes, Leslie. *The Suitcase Farming Frontier: A Study in the Historical Geography of the Central Great Plains*. Lincoln: University of Nebraska Press, 1973.

Hicks, L. E. "Storage of Storm-Waters on the Great Plains." *Science* 19, no. 478 (April 1, 1892): 183–184.

Hill, James H. "A History of Baca County." MA thesis, Colorado State College of Education, Greeley, Colorado, 1941.

Historic Context Study of the Purgatoire River Region. Denver: Colorado Preservation, Inc., Fall 2011.

Hobbs, Gregory J., Jr. *Citizen's Guide to Colorado Water Law*, 3rd ed. Denver: Colorado Foundation for Water Education, 2009.

Hobbs, Gregory J., Jr. "Colorado Water Law: An Historical Overview." *University of Denver Water Law Review* 1, no. 1 (Fall 1997): 1–74.

Hoggatt, Volney T. *How to Get a Homestead in the National Forest Reserves*. Denver: Denver Post, 1911.

Hundley, Norris, Jr. *The Great Thirst: Californians and Water—a History*, rev. ed. Berkeley: University of California Press, 2001.

Hurt, R. Douglas. "Federal Land Reclamation in the Dust Bowl." *Great Plains Quarterly* 6, no. 2 (Spring 1986): 94–106.

Jackson, Donald C. "Engineering in the Progressive Era: A New Look at Frederick Haynes Newell and the U.S. Reclamation Service." *Technology and Culture* 34, no. 3 (1993): 539–574. Accessed October 4, 2023. https://doi.org/10.1353/tech.1993.0045.

Jones, P. Andrew, and Tom Cech. *Colorado Water Law for Non-Lawyers*. Boulder: University Press of Colorado, 2009.

Keating, Edward. *The Gentleman from Colorado: A Memoir*. Denver: Sage Books, 1964.

Ketchum, Milo S., Jr. "Milo Smith Ketchum—Dean, 1904–1919." Accessed August 4, 2014. www.ketchum.org/milo.

Kluger, James R. *Turning on Water with a Shovel: The Career of Elwood Mead*. Albuquerque: University of New Mexico Press, 1992.

Kreiger, William E. "Geology and Petrology of the Two Buttes Intrusion." PhD dissertation, Pennsylvania State University, State College, 1976.

Larson, Robert W. *Populism in the Mountain West*. Albuquerque: University of New Mexico Press, 1986.

Larson, T. A. *History of Wyoming*, 2nd ed. rev. Lincoln: University of Nebraska Press, 1978.

Leonard, Stephen J., Thomas J. Noel, and Donald L. Walker Jr. *Honest John Shafroth: A Colorado Reformer*. Colorado History Series 8. Denver: Colorado Historical Society, 2003.

Limerick, Patricia Nelson. *Desert Passages: Encounters with the American Deserts*. Albuquerque: University of New Mexico Press, 1985.

Logan, Michael F. *The Lessening Stream: An Environmental History of the Santa Cruz River*. Tucson: University of Arizona Press, 2002.

Long, Clarence S. "Basal Cretaceous Strata, Southeastern Colorado." PhD dissertation, University of Colorado, Boulder, 1966.

Lovin, Hugh T. *Complexity in a Ditch: Bringing Water to the Idaho Desert*. Pullman: Washington State University Press, 2017.

Lovin, Hugh T. "The Farwell Trust Company of Chicago and Idaho Irrigation Finance." *Idaho Yesterdays* 38, no. 1 (Winter 1994): 7–17.

Lovin, Hugh T. "LaSalle Street Capitalists, Charles Hammett and Irrigated Farming at King Hill." *Pacific Northwest Quarterly* 98, no. 1 (Winter 2006–2007): 29–38.

Macdonell, Lawrence J. *From Reclamation to Sustainability: Water, Agriculture, and the Environment in the American West*. Niwot: University Press of Colorado, 1999.

MacKendrick, Donald A. "Before the Newlands Act: State Sponsored Reclamation Projects, 1888–1903." *Colorado Magazine* 52, no. 1 (Winter 1975): 1–21.

Mackey, Mike. *Henry A. Coffeen: A Life in Politics*. Powell, WY: Mackey, 2012.

MacKinnon, Anne. *Public Waters: Lessons from Wyoming for the American West*. Albuquerque: University of New Mexico Press, 2021.

Mahard, Richard H. "A History of the Department of Geology and Geography, Denison University, Granville." *Ohio Journal of Science* 79, no. 1 (January 1979): 18–21.

Martin, Helen. "A History of Water Resources and Rights in the Northern San Luis Valley." MA thesis, University of Northern Colorado, Greeley, 2003.

McCarthy, G. Michael. "Insurgency in Colorado: Elias Ammons and the Anticonservationist Impulse." *Colorado Magazine* 54, no. 1 (Winter 1977): 26–43.

McLain, Robert A. *Peopling the "Picketwire": A History of the Pinon Canyon Maneuver Site*. Champaign, IL: US Army Corps of Engineers, Engineering Research and Development Center, Construction Engineering Laboratory, July 2007.

Mead, Elwood. "Systematic Aid to Settlers in First Need." *Irrigation Age* 29, no. 7 (May 1914): 202–204, 216.

Mead, Elwood. *Irrigation Institutions*. New York: Macmillan, 1903.

Mehls, Steven F. "An Area the Size of Pennsylvania: David H. Moffat and the Opening of Northwest Colorado." *Midwest Review*, second series 7 (Spring 1985): 15–30.

Mehls, Steven F. "Westward from Denver: The Obsession of David Moffat." *Railroad History* 146 (Spring 1982): 29–40.

Meine, Curt. *Aldo Leopold: His Life and Work*. Madison: University of Wisconsin Press, 2010.

Meyers, Charlie. "Hot Spot during Dry Spell." May 8, 2016. www.denverpost.com/sports/ci_2797131#. Accessed October 23, 2020.

Moody's Manual of Railroads and Corporation Securities. Twenty-Second Annual Number, Industrial Section, vol. 2: K to Z. New York: Moody's Manual, 1921.

Moser, Clifford A. "Lake Setchfield: A Management Area for Wildlife in Southeastern Colorado." *Colorado Outdoors* 10, no. 6 (November–December 1961): 20–21.

Nostrand, Richard L. *The Hispano Homeland*. Norman: University of Oklahoma Press, 1992.

Official Proceedings of the Eighteenth National Irrigation Congress Held at Pueblo, Colorado, September 26–30, 1910. Pueblo, CO: Franklin Press, 1910.

Official Proceedings of the Eleventh National Irrigation Congress Held at Ogden, Utah, September 15–18, 1903. Ogden, UT: Proceedings Publishing Company, 1904.

Official Proceedings of the Fourteenth National Irrigation Congress Held at Boise, Idaho, September 3–8, 1906. Boise: Statesman Printing, 1906.

Official Proceedings of the Third National Irrigation Congress Held at Denver, Colorado, September 3–8, 1894. Denver: Committee of Arrangements, 1894.

Ogburn, Robert W. "A History of the Development of the San Luis Valley Water." *San Luis Valley Historian* 28, no. 1 (1996): 5–40.

Orsi, Jared. "From Horicon to Hamburgers and Back Again: Ecology, Ideology, and Wildlife Management, 1917–1935." *Environmental History Review* 18, no. 4 (1994): 19–40.

Osteen, Ike. *A Place Called Baca*. Chicago: Adams Press, 1979.

Peterson, Mark, Nathan Pieplow, and Andrew Spencer. "Birding in Colorado: Birds in

Baca County, Colorado." n.d. Accessed June 28, 2020. coloradocountybirding.org.

Philpott, William. *Vacationland: Tourism and Environment in the Colorado High Country*. Seattle: University of Washington Press, 2013.

Pisani, Donald J. "Enterprise and Equity: Critique of Western Water Law in the Nineteenth Century." *Western Historical Quarterly* 18, no. 1 (January 1987): 15–37.

Pisani, Donald J. *From the Family Farm to Agribusiness: The Irrigation Crusade in California and the West, 1850–1931*. Berkeley: University of California Press, 1984.

Pisani, Donald J. "Reclamation and Social Engineering in the Progressive Era." *Agricultural History* 57, no. 1 (January 1983): 46–63.

Pisani, Donald J. *To Reclaim a Divided West: Water, Law, and Policy, 1848–1902*. Albuquerque: University of New Mexico Press, 1992.

Pisani, Donald J. *Water and American Government: The Reclamation Bureau, National Water Policy, and the West, 1902–1935*. Berkeley: University of California Press, 2002.

Pisani, Donald J. *Water, Land, and Law in the West: The Limits of Public Policy, 1850–1920*. Lawrence: University Press of Kansas, 1996.

Poor's Manual of Industrials, Manufacturing, Mining and Miscellaneous Companies. New York: Poor's Manual, 1916.

Portrait and Biographical Record of Dubuque, Jones and Clayton Counties, Iowa. Chicago: Chapman, 1894.

Proceedings of the Conference of Western Governors Held at Denver, Colorado April 7, 8, 9, 10, 11, 1914. Denver: Smith-Brooks, 1914.

Proceedings of the Conference of Western Governors Held at Salt Lake City, Utah June 5, 6, and 7, 1913. Denver: Smith-Brooks, 1913.

Proceedings of the Trans-Missouri Dry Farming Congress, Held at Denver, Colorado, January 24, 25, 26, 1907. Denver: Denver Chamber of Commerce, 1907.

Quintana, Frances Leon, with Richard O. Clemmer, contributor. *Ordeal of Change: The Southern Utes and Their Neighbors*. Walnut Creek, CA: Altamira, 2004.

Raley, Brad F. "Private Irrigation in Colorado's Grand Valley." In *Fluid Arguments: Five Centuries of Western Water Conflict*, ed. Char Miller, 156–177. Tucson: University of Arizona Press, 2001.

Reiger, John F. *American Sportsmen and the Origins of Conservation*, 2nd ed. Corvallis: Oregon State University Press, 2001.

A Report on Sport Fish Restoration. Washington, DC: US Bureau of Sport Fisheries and Wildlife, 1957.

Sanderson, James, and Michael Moulton. *Wildlife Issues in a Changing World*, 2nd ed. Boca Raton: CRC Press, 1990.

Schorr, David. *The Colorado Doctrine: Water Rights, Corporations, and Distributive Justice on the American Frontier*. New Haven, CT: Yale University Press, 2012.

Schulp, Leonard. "I Am Not a Cuckoo Democrat! The Congressional Career of Henry A. Cofeen." *Wyoming Annals* 66, no. 3 (Fall 1994): 30–47.

Schulte, Steven C. *As Precious as Blood: The Western Slope in Colorado's Water Wars, 1900–1970*. Boulder: University Press of Colorado, 2016.

Schulte, Stephen C. *Wayne Aspinall and the Shaping of the American West*. Boulder: University Press of Colorado, 2002.

Seventeenth Annual Report of the Commission of Banking Being the Twentieth Annual Report of the Banking Department of the Commonwealth of Pennsylvania for the Year 1911, part I: *Banks, Savings Institutions and Trust Companies*. Harrisburg, PA: The Commonwealth, 1912.

Sheflin, Douglas. *Legacies of Dust: Land Use and Labor on the Colorado Plains*. Lincoln: University of Nebraska Press, 2019.

Sherow, James Earl. "Agricultural Marketplace Reform: T. C. Henry and the Irrigation Crusade in Colorado, 1870–1914." *Journal of the West* 31, no. 4 (October 1992): 51–58.

Sherow, James Earl. "The Chimerical Vision: Michael Creed Hinderlider and Progressive Engineering in Colorado." *Essays and Monographs in Colorado History* 9 (1989): 37–59.

Sherow, James E. *The Grasslands of the United States: An Environmental History*. Santa Barbara: ABC-CLIO, 2007.

Sherow, James Earl. "Utopia, Reality, and Irrigation: The Plight of the Fort Lyon Canal Company in the Arkansas River Valley." *Western Historical Quarterly* 20, no. 2 (May 1989): 162–184.

Sherow, James Earl. *Watering the Valley: Development along the High Plains Arkansas River, 1870–1950*. Lawrence: University Press of Kansas, 1990.

Simmons, Virginia McConnell. *The Ute Indians of Utah, Colorado, and New Mexico*. Boulder: University Press of Colorado, 2001.

Stiller, David. *Water and Agriculture in Colorado and the American West: First in Line for the Rio Grande*. Reno: University of Nevada Press, 2021.

Stone, Wilbur F. *History of Colorado*, vols. 1–5. Chicago: S. J. Clarke, 1918.

Sturgeon, Stephen C. "Just Add Water: Reclamation Projects and Development Fantasies in the Upper Basin of the Colorado River." *Library Faculty and Staff Publications* 62, 2008. Accessed January 16, 2016. http://digitalcommons.usu.edu/lib_pubs/62.

Sturgeon, Steven C. *The Politics of Western Water: The Congressional Career of Wayne Aspinall*. Tucson: University of Arizona Press, 2002.

Taylor, Morris F. "The Town Boom in Las Animas and Baca Counties." *Colorado Magazine* 55, nos. 2–3 (Spring–Summer 1978): 111–132.

Teele, R. P. "The Financing of Non-Governmental Irrigation Enterprises." *Journal of Land and Public Utility Economics* 2, no. 4 (October 1926): 427–440.

Teele, R. P. "Nettleton, Edwin S." In *Cyclopedia of American Agriculture*, vol. 4, ed. L. H. Bailey, 599. London: Macmillan, 1909.

Teisch, Jessica B. *Engineering Nature: Water, Development, and the Global Spread of American Environmental Expertise*. Chapel Hill: University of North Carolina Press, 2011.

Tyler, Daniel. *The Last Water Hole in the West: Colorado–Big Thompson Project and the Northern Water Conservancy District*. Boulder: University Press of Colorado, 1992.

Tyler, Daniel. *Silver Fox of the Rockies: Delphus E. Carpenter and Western Water Compacts*. Norman: University of Oklahoma Press, 2003.

Tyler, Daniel. *WD Farr: Cowboy in the Boardroom*. Norman: University of Oklahoma Press, 2011.

Underwood, Kathleen. *Town Building on the Colorado Frontier*. Albuquerque: University of New Mexico Press, 1987.

Walker, Roger. "The Delta Project: Utah's Successful Carey Act Project." 1985. Accessed October 24, 2022. wwww.waterhistory.org.

Warren, Edward Royal. "A Collecting Trip to Southeastern Colorado." *The Condor* 8, no. 1 (January–February 1906): 18–24.

Warren, Edward Royal. *The Mammals of Colorado*. New York: G. P. Putnam's Sons, 1910.

Warren, Hugh. "The History of Bent County, Colorado." MA thesis, Colorado State College of Education, Greeley, Colorado, 1939.

Watrous, Ansel. *History of Larimer County, Colorado*. Fort Collins: Courier Printing and Publishing, 1911.

Webb, Walter Prescott. *The Great Plains*. Boston: Ginn, 1931.

Weeks, Michael. *Cattle Beet Capital: Making Industrial Agriculture in Northern Colorado*. Lincoln: University of Nebraska Press, 2022.

Welch, Michael E. *A Mission in the Desert: Albuquerque District, 1935–1985*. US Army Corps of Engineers. Ann Arbor: University of Michigan Library, 1985.

Wheelock, David C. "Regulation and Bank Failures: New Evidence from Agricultural Collapse of the 1920s." *Journal of Economic History* 52, no. 4 (December 1992): 806–825.

White, Richard. *"It's Your Misfortune and None of My Own": A New History of the American West*. Norman: University of Oklahoma Press, 1991.

White, Richard. "Trashing the Trails." In *Trails to a New Western History*, ed. Patricia Nelson Limerick, Clyde A. Milner II, and Charles E. Rankin, 26–39. Lawrence: University Press of Kansas, 1991.

Whol, Ellen. *Wide Rivers Crossed: The South Platte and the Illinois of the American Prairie*. Boulder: University Press of Colorado, 2013.

Wickens, James F. *Colorado in the Great Depression*. New York: Garland, 1979.

Woeste, Victoria Sakes. *Henry Ford's War on Jews and the Legal Battle against Hate Speech*. Stanford: Stanford University Press, 2012.

Wolf, Tom. *Arthur Carhart: Wilderness Prophet*. Boulder: University Press of Colorado, 2008.

Worster, Donald. *Rivers of Empire: Water, Aridity, and the Growth of the American West*. New York: Pantheon Books, 1986.

Young, Richard Keith. *The Ute Indians of Colorado in the Twentieth Century*. Norman: University of Oklahoma Press, 1997.

INDEX

ABOUT THE AUTHOR

GERALD C. MORTON is an independent historian who specializes in business and environmental history. He holds degrees from Adams State University and Colorado State University.